W9-BSG-417

PERSPECTIVES ON ECOLOGY

Also by Koula Mellos
L'IDEOLOGIE ET LA REPRODUCTION DU CAPITAL
(*co-author*)

Perspectives on Ecology

A Critical Essay

Koula Mellos

Associate Professor
Department of Political Science
University of Ottawa

St. Martin's Press New York

First published in the United States of America in 1988

Printed in Hong Kong

ISBN 0–312–02417–7

Library of Congress Cataloging-in-Publication Data
Mellos, Koula, 1942-
Perspectives on ecology.
Bibliography: p.
Includes index.
1. Ecology—Philosophy. 2. Ecology. I. Title.
QH540.5.M45 1988 303 88–18183
ISBN 0–312–02417–7

Contents

Preface

One of the most critical issues of our time is the issue of ecology. There is concern about harmful environmental and ecological effects of multifaceted and rampant pollution, acid rain, radioactive environmental emissions of nuclear technology, depletion of some national resources, extinction of some animal and plant species and dramatic genetic mutation of others. Some people regard these as being of such critical proportion as to constitute an ecological crisis not only of quality of life but of survival. Most ecologically sensitive people agree that human activity is the source of the current ills and agree also by the same token that changes in human activity are necessary to restore ecological harmony.

The relationship between human activity and the state of ecology motivates much of contemporary ecological-social theory both as analysis of social sources of ecological disturbances and as normative statements of necessary corrective action. Socio-ecological theory is not, however, a cohesive body of thought. It takes different forms, each drawing a particular and differing analysis of the issues and differing proposals. This essay is a modest attempt to identify some principal currents of socio-ecological theory and to assess critically their respective analyses of the ecological crisis and their proposals for change.

In working out the problems raised in this essay I benefited greatly from discussions with friends and colleagues. I should like to thank Léonard Beauline, Raymundo de Andrade, Audun Øfsti, Douglas Moggach and especially Roberto Miguelez.

This work was initially undertaken with the financial assistance of the Faculty of Social Sciences, University of Ottawa, and completed with the secretarial assistance of Ginette Rozon, Louise Clément and Phuong Chi Hoang of the research secretariat to whom I am grateful.

KOULA MELLOS

Preface

One of the most critical issues of our time is the issue of ecology. There is concern about harmful environmental and ecological effects of unmitigated and rampant pollution, acid rain, radioactive environmental emissions of nuclear technology, depletion of some natural resources, extinction of some animal and plant species and dramatic genetic diminution of other. Some people regard these as being of such critical proportions as to constitute an ecological crisis not only of quality of life but of survival. Most ecologically sensitive people agree that human activity is the source of the current ills and agree also by the same token that changes in human activity are necessary to restore ecological harmony.

The relationship between human activity and the state of ecology occupies much of contemporary ecological-social theory, both as analysis of social sources of ecological disturbances and as normative statements of necessary corrective action. Socio-ecological theory is not, however, a cohesive body of thought. It takes different forms, each drawing a particular and differing analysis of the issues and differing proposals. This essay is a modest attempt to identify some principal currents of socio-ecological thought and to assess critically their respective analyses of the ecological crisis and their proposals for change.

In working out the problems raised in this essay I benefited greatly from discussions with friends and colleagues. I should like to thank Leonora Beaulne, Raymundo de Andrade, Ardath Olsa, Douglas Moggach and especially Roberto Albuquerque.

This work was initially undertaken with the financial assistance of the Faculty of Social Sciences, University of Ottawa, and completed with the secretarial assistance of Charlotte Boizet, Louise Alemani and Phuong Chi Hoang of the research secretariat to whom I am grateful.

Alan Atkinson

Introduction

Ecology in the Oxford dictionary is defined as a science of the economy of animals and plants or that branch of biology which deals with the relations of living organisms to their surroundings, their habitats and modes of life. In this sense ecology (from the Greek *oikos* meaning house or habitat and *logos* meaning science) defines its object as constitutive of physical, chemical and biological processes and seeks to discover the organic laws governing such processes.

In this conception, disturbances or interferences in the normal reproduction of eco-systems are accountable or explainable on the physical, chemical and biological levels whether these disturbances emanate from the socio-cultural level of reality or whether they are generated on these same levels as, for example, a shift in the position of the earth's axis or a genetic mutation. In the perspective of traditional ecology as a natural science, the effects of operations of the socio-cultural level on eco-systems are reducible to the laws governing organic matter of a physical, chemical or biological nature.

In the last twenty years or so another meaning has been attached to the term 'ecology'. This meaning of ecology refers to the quality of the environment and alludes to a concern about a perceived condition of environmental deterioration. In this popular sense, ecologists are not dispassionate scientists claiming value neutrality in their study of eco-systems. They are ecologists expressing a concern for the bio-physical state of the environment, searching for causes of perceived environmental damages and disequilibria, acting in ways which aim at correcting the environmental problems numerous as they are considered to be. Amongst environmental conditions which are raised to the level of salient issues are depletion of natural resources, air pollution, chemical contamination of soil and water, acid rain, and radiation. There is some belief that the life-supporting capacity of the planet's air, water and soil is rapidly waning as these resources become contaminated with toxic industrial emissions threatening interconnected life cycles of total ecological systems and human life itself. There is also a belief that radio-activity defies containment and constitutes the immediate killer of life.[1] All these are considered to be a threat in varying degrees not only to non-human eco-systems but to human life in its biological and social forms.

This meaning of ecology emerged in a practice different from that of scientific research and theorising in which emerged the former sense of ecology. It emerged in a practice of some form of protest or opposition to that economic activity and those political decisions whose effects appear to be harmful in some way to human life and to the environment. It alludes, therefore, to a social sensibility, however limited or extensive, to the quality of the environment and to the quality of life which such an environment allows. Of course, social sensibility to environmental questions is not purely a modern phenomenon. Environmentalism in this popular sense, has a far longer history. In the period prior to the '50s and '60s it focused predominantly on conservation of particular animal species whose population sizes appeared to be decreasing and whose reproduction as a species seemed to be endangered. Conservation was the objective of various groups such as the Sierra Club, the Audubon Society and the Isaak Walton League. To this day these groups call for the preservation, creation or extension of national parks in which areas are preserved as sanctuaries for wild life. The rather recently formed environmentalist group, Greenpeace, which has concentrated some of its energy on programmes for the conservation of the whale and baby seal, follows this tradition.

The social sensibility to which the term ecology is now linked traces the quality of life and of the environment to social decisions and social activity. It is a social sensibility for creating a harmony between social activity and the environment or at least reducing the environmentally negative effects of this activity. This notion of ecology thus encompasses both organic nature and social relations. Each affects the other and thus the state of the environment is in part socially determined. This form of social sensibility about the environment is that to which the term 'ecological movement' of the last twenty years, or so, refers.

This socio-ecological sensibility or ecologico-social movement takes many practical forms – ideological, political and economic. In its ideological form, it is a struggle to promote a conception of reality as an interdependent whole including society and organic nature and in particular, a conception of society as one having positive effects on organic nature in the sense of being compatible with organic exigencies. It is a political struggle for creation of political relations which promote socio-ecological harmony. Some aspects of state intervention in the economy in regard to the economic effects on the environment are not totally unrelated to this political struggle. The relatively newly-formed ministries of the environment as regulating bodies in most western countries, with the function of monitoring industrial activity and controlling its toxic emissions into the environment, grew out of environmentalist concerns, however ineffective the state's performance in ths regard is deemed to be by ecology activists. More recent evidence of this political struggle is the emergence of 'Green' political parties in some western countries such as West Germany and France with a general overall programme for an ecologically sensitive social reconstruction. More militant forms of action such as demonstrations and sit-ins centring on the issue of nuclear energy and calling for an end to nuclear weapons production, the arms race and war, are increasingly in evidence in these same countries. Some call for an end to the use of nuclear energy altogether.[2]

The socio-ecological movement also has an economic form as a struggle for individual self-sufficiency. Co-operatives and citizens' groups organise in rural or urban areas to produce means of livelihood in line with ecological technologies such as recycling, organic farming and solar energy. Some groups such as the 'Farm', which was first established in Tennessee in 1971 and now has branches in other areas of the United States and some parts of Canada, aim at developing ecologically sound means to achieve total self-sufficiency including self-education and self-health care.

The notion of ecology or the new ecology has emerged in these struggles and it is these struggles to which the term 'ecological movement' is generally understood to refer. This social movement however, cannot be described as being homogeneous either in ideological, political or economic practice. It covers disparate groups in regard to a conception of ecological society and to ideological, political and economic practice for achieving such a society. It includes pacifist and anti-nuclear groups who concentrate on exerting pressure on the state to reduce or eliminate the production and deployment of nuclear armaments and nuclear energy for industrial purposes. It includes nutritional groups who propose the replacement of modern medicine and present forms of food production and distribution by preventive mental and physical health practices and organic food production and cooperative distribution. It includes the entire alternative movement including food cooperatives, urban communes, 'soft' energy cooperatives, 'appropriate' shelter advocates who resist the planned obsolescence and material wastage of capitalist market relations. Such groups resist consumer dependence by developing alternative means of life involving 'ecotechnics' and self-sufficiency.

The pacifist, anti-nuclear groups, the nutritional groups, the alternative groups constitute a more coherent movement in ideological, political and economic struggles. The last two in particular, in waging an economic struggle for liberation from capitalist dependence, also form a cultural movement for alternative life styles and self-sufficiency as a political instrument of social transformation.[3] The differences between the more traditional, reformist environmentalists on the one hand, and the more radical anti-consumerist alternative ecologists on the other, often translate into intergroup conflict such as can be noted, for example, in the critique which the latter levies at the former for being reformist and ultimately compatible with existing organisations.[4]

Whatever difference and affinities between these various distinct groups, from the more social democratic reformist environmentalists, and 'alert consumers' to the more radical alternative ecologists, they are not of equal strength, support and resources in all western countries. The anti-nuclear groups are stronger in Germany and England and the alternative groups are extensive in the United States.

The ecology movement is not a mass movement either in a quantitative or sociological sense. Although the anti-nuclear movement has gained some 'popular' support in its opposition to nuclear arms and war, and in this unites not only traditional environmentalists and radical ecologists, but also people from the left and other pacifists, the ecology movement in general is largely an urban middle-class movement. This class basis is not of negligible significance, as we examine below, in its critique of contemporary society both capitalist and socialist and in its concept of a just society. Some of the theoretical inspiration of this movement is the Frankfurt School, particularly Marcuse, who influenced much of the counter-culture movement of the '60s – a movement which has been absorbed by or has evolved into the ecology movement, feminism and gay liberation. The basic theoretical affinity between these various forms of social protest is their theory of domination, both capitalist and socialist, which claims that a relation of control characterises social relations as well as the essentially utilitarian relation with nature. It is in the alternative movement that the critique of domination is most extensive and most developed[5] particularly in North America. The socio-ecological self-management movement in France shares many basic conceptions of contemporary society with that of the North American alternative movement as is evidenced in their respective theories of social change.[6]

All groups within the ecology movement share a conception of interconnectedness of society and organic nature although they may not totally agree on the character of this interconnectedness. If this ontological axiom, however, is taken as the point of reference to the various groups of the ecology movement or submovements within the general ecology movement and indeed if this is taken as the first distinguishing feature between the new ecology and the traditional – the new ecologists and traditional conservationists – it does not apply simply to the social protest movements. The relationship between society or the socio-cultural level of reality and the biological, chemical and physical levels of organic nature is raised in the form of an ecological problematic by groups which are not normally considered to be part of the ecology movement.

A school of neo-Malthusian thinking emerged in the post Second World War period and culminated in the founding of the Club of Rome in the late '60s. It comprises eminent social scientists, including economists, social demographers, urban planners, etc. and natural scientists who claim that in the process of social expansion ecological imbalances have occurred in the form of, for example, natural resource exhaustion and pollution of near-irreversible proportions. The solutions they propose for ecological restoration are social solutions just as are those of radical ecologists, however different in form they may be from the latter. Indeed, they propose a reduction of various forms of growth including demographic expansion, which they claim occurs at exponential rates, to rates compatible with their conception of planetary finitude and life-supporting capacity. While the ecological sensibility of this group takes a distant form both in theory and in practice from that of the various groups within the ecological movement as we have defined it here and as is normally understood, it participates in the ecological debate and has a far from negligible effect on levels of political and economic decision-making affecting the whole society.

Another ecologically sensitive school of thought not normally considered to constitute part of the ecology movement is the school of eco-development. It is shared by economists, sociologists affiliated with international organisations such as the United Nations and some of its organisms specifically involved with patterns of economic development in the Third World. It too contends that a relationship of interconnectedness exists between society and organic nature and that consequently particular patterns of economic development will have determinate effects on organic nature. This socio-ecological theory adopts a human needs approach which posits the nature of man as constituting basic needs best satisfied by a pattern of production and distribution based on moderation and equality, decentralisation and self-determination and means of production or technology suited to the particular socio-ecological conditions of a given community and ecological area of production.

If a posited interconnectedness of society and nature allows the inclusion of neo-Malthusianism and eco-development in the socio-ecological phenomenon, it also covers a school of thought on ecology which, rather than joining ecologists in their concerns, opposes the whole movement including the neo-Malthusian and eco-development and thus can hardly be considered to be ecologically sensitive in the sense of the others. Indeed, its theory of economic development paints a rather optimistic view of the effects of society on nature. This theory, however, confronts ecology explicitly and thereby enters the ecology debate bringing to this debate a form of ecological sensitivity which some would prefer to characterise otherwise. This is highlighted in the contention that any ecological disturbances which have occurred or may occur in the future can be conquered by technological means and this requires a continued commitment to the present form of economic development which is a necessary stimulus for technological innovations. In this view, it is the unlimited technical mastery over nature provided by expanding technology which is the key to the solution of ecological problems as they emerge, however wide their magnitude may be. Advocates of this position are members and sympathisers of the Hudson Institute such as the American futurologists Herman Kahn and Gail Potter, British physicist and futurologist John Maddox, British economist Wilfred Beckerman, while advocates of eco-Malthusianism are members or sympathisers of the Club of Rome such as Aurelio Peccei, Dennis Meadows, Paul Ehrlich, Garrett Hardin, Kenneth Boulding, William Ophuls, and advocates of eco-development are Johann Galtung and Ignacy Sachs among others.

These forms of ecological sensibility are clearly widely disparate particularly if expansionism is included amongst them. These differences which are so evident in regard to the effect of social relations on organic nature derive from more basic and fundamental assumptions about the nature of reality. As distinct theories about nature's mediation of social reproduction, each claims to be valid over all others. The claim to validity is often defended or promoted by a contention of invalidity in competing claims as, for example, in the critique of expansionism. They are, in a strong sense, conflicting and competing theories of society or, more broadly, theories of society and organic nature as a totality.

This conflict on the level of theoretical discourse is not unrelated to the social struggles at the political and economic levels either as determinant of these struggles or as determined by them. A theoretical proposition of a materialist nature and one of critical importance underlies this statement however true this statement appears to be strictly on empirical grounds. It is that ideas, attitudes and beliefs are generated in and have an effect on concrete social relations and by the same token a theory of society and of its relation to organic nature is not exempt from such a social determination nor is it of no consequence on concrete social relations about which it is a theorised construction.

If we look at any of these theories, even in a perfunctory manner, we quickly see that none advocates a detachment from social action. Each theory, on the contrary, is a call to action of various forms – political, economic, cultural even though some of these theories (neo-Malthusian and expansionist, in particular) claim validity in the name of science and its tenet of value–neutrality. The notions and understanding each theory conveys are not intended to remain purely on the conceptual level but to convert into practice by orienting and mobilising action such that social relations are preserved, modified, or radically altered in a way favourable to the normative character of the theory. These are competing theories in the sense of implying not simply conflicting action in the pluralist sense of co-existing differences, but a struggle for the preservation, reform or radical alteration of society as a whole, including its relation with nature. They are not only competing validity claims but also, and more importantly for social theory, competing legitimacy claims.

Because each theory is in a relation of struggle competing for validity and legitimacy with other theories, the full significance of those theories normally thought of as being part of the tradition of critical protest can be understood when examined against the others not normally linked to the ecology movement. Such a comprehensive focus of the various positions constituting the current ecological debate may serve to reveal not simply the antagonistic differences between them. A comprehensive focus on these theories in their relation of struggle serves also to put into methodological and theoretical perspective the question of determinants of these theories. The question of varying reconstitution of the same concrete socio-natural reality in theory relates to the social structure as a perspective from which the socio-natural reality is reconstituted in theory. This approach helps to focus on the particular mediations or determinants of theoretical construction in which the same object of concrete reality, namely ecology, becomes radicalised in one theory while kept unmodified in another. It calls attention to the structure of social relations in which one and the same object is constituted as a radical issue, issue of reform and non-issue at the same time but from different perspectives of the social structure. It calls attention to the struggle for legitimacy including the translation of the theory into practice.

In search of these social determinants of theoretical ecology or in other words, of the variety of theoretical reconstitution of ecology, this study retains at least one form (expansionism) which would not normally be considered to address itself to ecology, and forms (neo-Malthusianism and eco-development) not considered a part of the radical ecology protest movement. The study we propose to conduct in the pages that follow is situated on two different levels simultaneously: (a) the level of theoretical construction of each theory and (b) the level of social determination of each theory. In order to be able to say anything about a theory we must know how the theory depicts itself and in order to know how the theory depicts itself we must know something of the basis of its determination. In analytically reconstructing a theory we are never on a purely descriptive level. We are never solely 'inside' the theory even though our first task is to faithfully capture the intended sense of the theory for it is this which allows us to explain its intentions. A theory, however comprehensive it may be, refers, paradoxically perhaps, to something beyond itself. It is this 'beyond itself' including its social determination which a meta-theoretical critique reveals. To this end, we shall examine the theory of reality including, and in particular, the relation between the socio-cultural level and organic level which each theory, as a system of representation of the concrete world, advances. This is a difficult task for this relation is not always specified in these theories. Most of these socio-ecological theories are not articulated on a level which highlights ontological notions. This is particularly so of neo-Malthusianism and expansionism which are expressed largely in the form of empirical studies of patterns and tendencies and projections of various forms of expansion such as demographic, capital and their interrelations and global effects. The neo-Malthusian, *The Limits to Growth*, authored by Donella Meadows *et al.*, the first Report to the Club of Rome, and the expansionist, *The Year 2,000: A Framework for Speculation on the Next Thirty-Three Years*, are examples of this approach by neo-Malthusians and expansionists respectively. To a large extent an empirical methodological approach characterises the school of eco-development too, although, of course, the chosen models of development differ markedly from neo-Malthusianism and expansionism. It is, therefore, a difficult task to identify ontological assumptions in these cases. It is not, however, impossible for whether or not assumptions of an ontological nature are expressed; whether particular values are expressed or explicitly denied in favour of the value-neutrality premise of empirico-analytical science, they are present in some implicit form or at some more fundamental level which may often escape the awareness of the author. Our task is to reconstruct the theory paying as much heed to the 'unstated' as to the explicitly claimed especially in the form of the normative claims in order to interpret the particular socio-ecological sense of reality in each. Some texts are more 'revealing' of assumptions and values than others and it is on these that we shall concentrate our attention.

If our purpose is to uncover the basic assumptions and values towards which members of a school of thought gravitate, and if these are more retrievable in some textual expressions than others, we may be well advised to concentrate on the more revealing texts. Accordingly, for a reconstruction of neo-Malthusianism we will focus on some writings of Aurelio Peccei, Dennis Meadows, Paul Ehrlich, Garret Hardin, Kenneth Boulding, Herman Daly, William Ophuls and others. Similarly, for expansionism we will concentrate on some writings of Herman Kahn, John Maddox, and Wilfred Beckerman, and for the thought of eco-development on the writings of Ignacy Sachs and Johann Galtung.

Radical ecology differs from neo-Malthusianism, expansionism and eco-development not only as regards substantive theory but as regards the level at which this theory is articulated. Radical ecology is posed at a certain ontological level, assuming a rather comprehensive form at least in the work of some prominent writers and activists such as Murray Bookchin in particular but also in André Gorz and William Leiss. Murray Bookchin claims an anarchist perspective and André Gorz his own particular version of Marxism but although there are distinct differences between them as there are between any number of radical ecologists, a comprehensive analysis of their thought reveals a basic affinity – one which unites other radical ecologists who in so many respects appear disparate and conflicting.

The general hypothesis which we should like to advance in this work is that an anthropological theory of the nature of man underlies each of the distinct theories of ecology and becomes not only its principal internal determinant but constitutes the key referent to the perspective of the social structure as the basis of determination. We propose the specific hypothesis that a common anthropology underlies neo-Malthusianism and expansionism and that this anthropology is a liberal anthropology. The basic premise of these theories is that a basic nature inheres in human beings and that it is possessive individualism. The fundamental difference between neo-Malthusianism and expansionism consists in their markedly differing response to this premise. Malthusianism is critical of this conception of human nature without however denying it an ontological status. Rather, it proposes a politically sanctioned restraint on individual activity commensurate with ecological tolerance. On the other hand, expansionism embraces this conception of human nature unaltered from its classical liberal form in which the uncontained activity of private appropriation is seen as the expression of human essence and as the means of correcting any possible ecological disturbances.

We also propose the hypothesis that a common anthropology underlies the more radical forms of ecology and in a more ambiguous form in eco-development as well but in most of these it is not a liberal anthropology but rather an anthropology of self-sufficiency. These theories assume that human nature is the individual positing of objectives and the development of means of achieving them without social mediation but rather in direct relation with organic nature. It is an anthropology of radical independence as asocial self-sufficiency. This hypothesis suggests that radical ecology is a form of organic reductionism where the organic laws of nature prevail in sustaining ecological equilibria since asocial self-sufficiency, essentially organic causality, becomes the form of self-realisation.

The identification of the distinct assumptions of human nature helps to account for the differences between neo-Malthusianism and expansionism on the one hand, and eco-development and radical ecology on the other, as well as the internal coherence which each distinct discourse displays. But, as we mentioned above in passing, it does more than this. It points beyond itself, to the specific social interests which each theory promotes, for a specific anthropology is also an interest structure. In pointing to these particular interests it relates to the perspective of a particular class in the class structure. If our hypotheses are correct, expansionism, neo-Malthusianism are theories of social relations in which the bourgeois class retains its relation of dominance *vis-à-vis* other classes. They are in this sense bourgeois ideologies and conceived from the perspective of the bourgeois class.

Radical theories of ecology, three of which we examine here, are theories which purport to change social relations such that the bourgeois class loses its dominance and other classes are liberated from domination by an immediate relation between the individual and his means of production – a relation of individual self-sufficiency. Radical ecology, just as expansionism and neo-Malthusianism, does not escape class determination. It is not, however, a bourgeois response to contradications of advanced capitalist relations. Neither is it a working-class response to these same contradictions. We suggest the hypothesis that it is articulated from the perspective of the petit bourgeois class expressing the contradictions and ambiguities of a class in a social structure in which the dominant class relation is that between the capitalist class and the working class. Radical ecology as 'asociality' is determined by the petit bourgeois class position in relation to the dominant capital–labour relation. To sustain this proposition we shall argue the sense in which the petit bourgeois class is both removed from and integral to the dominant relation such that this class is neither dominant as is the bourgeois class nor productive of surplus-value as is the working class but nevertheless necessary for the reproduction of this capitalist dominance.

Our assessment of the petit bourgeois class programme of liberation from domination suggests that it is largely a programme of detachment or escape from this dominant capitalist relation rather than of change of this dominant relation itself. It suggests further that the change of the dominant capital–labour relation must be generated from within this relation itself and hence necessarily involve the working class in a radical form of action which combines the advances proposed by radical ecology in a united action of dominated classes and which converts asociality into collective liberation and collective self-sufficiency.

Part I

Neo-Malthusianism and Expansionism

1 Neo-Malthusian Theory

In the late 18th century, Thomas Malthus, an English political economist, advanced a theory of crisis in his *Essay on the Principle of Population*,[1] based on a posited relation of disproportion between the rate of demographic growth and the rate of growth of food supply. According to this thesis, population naturally increases in geometric ratio but the means of subsistence, or agricultural production increases only in an arithmetic ratio making it impossible for agricultural production to sustain growing populations indefinitely. These two opposing natural tendencies generate periodic crises of food supply corrected by reduction of population size. Malthus describes two distinct forms of checks on population size: 'positive' checks such as war, epidemics, famine, and 'preventive' checks such as various forms of birth control, including abortion, and infanticide. Since food scarcity, however, is the condition for the operation of these checks, it is the ultimate check on population increase.

Malthus revised his population thesis somewhat in a later version of the *Essay*[2] by adding an additional type of population check. It is the check of 'moral restraint' from sexual intercourse such as would result from the postponement of age of marriage conditional to pre-marital sexual abstinence.

Both versions retain the fundamental premise of two opposing natural tendencies: unlimited demographic expansion on the one hand, and limited food production on the other. Neo-Malthusian theory shares this premise of opposing tendencies but applies it not only to a disproportionate relation of expansion between population and food production but also to a disproportion between rates of technological expansion, consumption of mineral resources, generation of various forms of pollution, on the one hand, and the planet's finite capacities, on the other. They include the earth's limited mineral and fossil reserves and its limited capacity to absorb pollutants. The notion of crisis encompasses thus more than demographic growth. It includes also technological and industrial expansion approaching or reaching the planet's 'carrying' capacity. It is a notion related to 'ecological scarcity' as a measure of natural organic tolerance.

This general premise of opposing tendencies marks the work of a great many thinkers of different scientific backgrounds including biology, geology, demography, economics, political science and sociology. Contributions to neo-Malthusian theory are thus disparate in regard to their focus and their methodological approach. They range from empirical studies of birthrates in a given country and time period, for example, to broad theoretical statements of social and political relations. Whatever their scope they have direct implications for social theory. That is, the relations that are drawn between population, technology and industrial expansion on the one hand and natural organic finitude on the other are relations meaningful within a particular theoretical problematic and this in turn is revealing of the particular normative assumptions of society and nature of neo-Malthusian theory in general.

The theoretical problematic of neo-Malthusianism as a framework which solicits and orients questions about society and nature is one which differs from the theoretical problematic of most forms of social theory. In this latter the problematic is one of social reproduction of social relations in relation to organic nature or, in other words, the material reproduction of social relations as an articulation of productive forces and productive relations. The neo-Malthusian problematic of social reproduction shifts from social structure to social size in purely quantitative terms of population size, size of technology, size of industrial infrastructure and their rates of quantitative expansion. The question of the relation between the social and the natural is posed in terms of the relation between social size and organic tolerance.

The conceptualisation of the problematic of social reproduction in these terms carries the implication that the fundamental problem of societies is not structurally determined social conflict but rather the finite organic capabilities of nature on which any form of life, social or biological, is dependent. In this, it takes up one of the axioms of natural science, namely that the natural system is one of finite dimensions with a determinate rhythm and pace of natural reproduction, and poses it as a condition of social reproduction. In this way, the problematic of social reproduction converts into a problematic of consumption of nature, nature dictating the limits of consumption commensurate with its own organically determined capabilities and tolerance. Nature's tolerance becomes the measure for determining acceptable population size, type of technology, size of industrial infrastructure, and the problem of societies becomes largely one of containing demographic, technological and industrial growth within these limits.

The relation between consumption of nature as determined by population size, exigencies of technology and industry and natural tolerance suggests the notion of ecology in this perspective. Ecology refers to the state of human consumption of nature's resoures – an equilibrium being defined as an extent and rate of consumption of resources including the generation of population proportional to the inherent capabilities of natural operations and a disequilibrium by a disproportion between consumption and natural tolerance. Those which determine the state of ecology are the rate of demographic growth and the rate of technological and industrial growth. A theory of crisis is derived from these theoretical constructions as overconsumption and pollution of nature by industries and populations of sizes exceeding nature's tolerance.

If this depiction captures the specificity of the neo-Malthusian theoretical problematic it casts some light on its premise of opposing tendencies. The problematic reveals a relation of contradiction between on the one hand exponential growth of population, technology and capital and on the other hand natural finitude but it leaves indeterminate the relation between the various factors of the contradiction. That is, the relation between rates of growth of population, technology and capital is left indeterminate and this indeterminateness or lack of articulation of the various components of the contradiction is a necessary logical consequence of a problematic which poses relations as relations of quantity. This is an important point in understanding the disparity and range of neo-Malthusian thinking. Some focus exclusively on population as does Malthus but this exclusive focus does not necessarily imply that population dynamics determine technological and industrial growth for the properties of population growth are theorized independent of technical and industrial growth. Paul Ehrlich,[3] Garrett Hardin,[4] Georg Borgstrom[5] are amongst those who single out the rate of population growth as naturally exponential and if not consciously controlled inexorably destructive of the entire planet. Others such as the MIT Team which authored the First Report to the Club of Rome,[6] *The Limits to Growth*, list a number of factors which increase exponentially namely population growth, capital growth, pollution growth and consumption of fossil mineral resources which reach an ultimate limit in view of natural finitude. *The Limits to Growth*[7] proposes a model for relating the dynamics of various forms of growth to the limits of growth as determined by natural finitude, as well as the various forms of growth to each other. The model, then, attempts to link the factors constitutive of the contradiction to each other but the relationships drawn are purely quantitative.[8] A brief look at this study may suffice to illustrate the sense of the quantitative character of the relations.

The study examines the relationship between five factors of growth, namely, population, capital, food production, production of non-renewable resources and production of pollution in the light of the planet's capacities and in the light of the impact of technological developments on this capacity. These relationships are cast in a dynamic context as rates of growth. The pattern of expansion of each factor in a given time period is constructed from extensive international data; they are related to each other and, as a whole process of growth, they are individually related to the 'carrying capacity' of the planet. The 'carrying' capacity refers to the planet's stock of resources and to its capacity to absorb pollution as determined by estimates of arable land, fresh water sources, fossil and mineral reserves and extent of pollution absorption.

The results which the Forrester system model yields from the combination of mathematical data of each of these factors of growth and carrying capacity of the planet is a disproportion between, on the one hand, rates of growth of population[9] and capital, rates of exploitation of fossil and mineral reserves, rates of production of pollutants and on the other hand, the available physical resources of the planet and their rate of renewal. Growth of population and capital is exponential and these rates of growth, particularly the exponential capital growth, generate exponential growth of fossil and mineral production and production of pollution. In view of the estimated finitude of the earth's capacities as regards arable land and water for food production, actual fossil and mineral reserves, capacity to recycle wastes and asborb pollutants, exponential growth is rapidly approaching the ultimate limits of physical carrying capacity. The model thus suggests that the factors of growth are interrelated such that exponential growth as a whole becomes a threat to the planet's survival. For example, the exponential growth of population generates an exponential demand for food but the supply of food is dependent on land, water and agricultural capital which depends in turn on capital growth and non-renewable resources of which there is a finite stock. Technical development is considered to have no appreciable impact on this contradiction in that it cannot extend sufficiently the earth's carrying capacity and hence the study predicts a planetary collapse by the year 2000 if growth proceeds exponentially.

The study has received a great deal of critique on the choice and use of mathematical data as measurements of growth and as measurements of the planet's physical capacities and resources. *The Limits of Growth* in its very quantitative methodology is revealing of the philosophy of neo-Malthusianism. It represents the spirit not only of the Club of Rome but of the neo-Malthusian school in general. There may be disagreement on estimates of a given factor on the mode of measuring estimates, or on the relative strength of one factor over others in generating a crisis. The more fundamental agreement of neo-Malthusians, however, is captured and highlighted in *The Limits to Growth*, namely that the dynamics of growth specific to populations and capital is exponential and this triggers off patterns of demand and consumption of food and natural resources which expand also exponentially and generate in the process an exponential production of pollution.

Such a dynamic of expansion of population and capital is posed as being determined by natural forces. It is conceived as being in the 'nature of things'. The entire normative orientation of neo-Malthusianism rests on this assumption. If it is inherent in the growth dynamics of populations and capital to expand exponentially and if this involves taking a magnitude of natural resources from nature at a pace beyond its physical capability to provide and giving back to nature wastes of a magnitude and at a pace exceeding the absorption capability of nature, human reason must intervene in this process of expansion to contain it to degrees compatible with natural tolerance. A steady state, equilibrium state, zero-growth state are proposed as the solution to the naturally determined contradiction. These are different terms which relate essentially to a single form, namely a state of compatibility between natural finitude and social reproduction. Before we turn to the social and political theory of contained growth, let us (a) examine the notion of capital growth in *Limits to Growth* as it is instructive of the underlying premise of the theory of contained growth which we shall argue is an anthropological premise of human nature of the liberal Hobbesian form, (b) look at one other neo-Malthusian theory of growth, that of William Ophuls,[10] for a sense of the similarity between different neo-Malthusian versions of the ecological crisis.

The Limits to Growth focuses on production of means of

consumption and means of production in order to calculate the rate of expansion of capital. The intent is thus to obtain a measurement of the rate of production of manufactured output for which it is supposed that the factor of labour can be kept methodologically constant in the simple assumption that it is sufficient.

> With a given amount of industrial capital (factories, trucks, tools, machines, etc.), a certain amount of manufactured output each year is possible. The output actually produced is also dependent on labor, raw materials, and other inputs. For the moment we will assume that these other inputs are sufficient, so that capital is the limiting factor in production.[11]

Capital is defined as the means of production as distinct from the means of consumption as the latter simply leaves the process of production. This capital stock expands by investments and this capital creates more output 'some variable fraction of the output is investment, and more investment means more capital. The new, larger capital stock generates even more output, and so on'.[12] This capital growth is regarded as being comparable to the process of expansion of population not only with respect to expansion but retraction. That is, just as the normal process of population expansion comprises a birth rate and a death rate in a relation of disequilibrium so does the normal process of capital expansion involve addition to capital stock at a rate higher than the rate of depreciation of capital stock such that the rate of 'world industrial capital stock in growing exponentially'.[13] It is here that the exponential growth of capital generates an exponential expansion of production of fossil fuels and mineral reserves as raw materials in the production process, which in turn generates pollution at an exponential rate of growth.

The Limits to Growth hastens to add qualifications to this depiction of the productive process of a kind which would imply that the productive process is a social and economic process and thus dependent on the socio-economic forces of a given social formation.

> Let us recognize, however, that the growth rates listed above are the products of a complicated social and economic system that is essentially stable and that is likely to change slowly rather than quickly, except in cases of severe social disruption.[14]

But this qualification does not alter or conceal the methodological significance of the drawn analogy between population growth and economic production. Economic production, however social it is considered to be, is seen as being governed by an inherent natural motion comparable to the biological motion of population expansion. Both are social processes but they are treated methodologically as if they were governed by natural laws. Not only is the analogy between population growth and capital growth revealing of this, but so is the method of abstraction from the social of rates of birth and death and rates of investment/depreciation as pure measures of quantity and socially invariable in ways other than a mere containment, or a reduction of these rates. In other words, reproduction is conceived on the model of biological reproduction and not on the model of social reproduction as a relation between productive forces and productive social relations in which the latter have primacy. We shall return to this point in our analysis of the social and political theory of neo-Malthusianism below.

William Ophuls's *Ecology and the Politics of Scarcity* is one of the most comprehensive and coherent statements of neo-Malthusian theory. It poses the relationship between human activity and physical substructure as a relationship of contradiction in the terms in which *The Limits to Growth* relates factors of growth to natural finitude. Unrestrained biological reproductive activity and economic activity proceeds at a pace which generates exponential growth making demands on the physical substructure which exceed the possibility to provide in view of the finitude of the physical environment. The unlimited demands of activity are confronted with a limited natural system. The social and political theory which follows is a theory of restraint of human activity to degrees compatible with physical finitude.

The originality of this work lies not so much in the analysis of patterns of growth and ecological limits but in the theory of containment of activity as a solution to the contradiction between unlimited demands and ecological scarcity. The analysis of the ecological effects of patterns of growth are nevertheless revealing in the similarity with *The Limits to Growth*. This similarity is not so much methodological as it is theoretical. Methodologically it is distinguished from the MIT research which applies the Forrester system model to comprehensive data in order to compute the pattern of expansion of a number of stipulated factors, principally population and capital. It is rather an analysis of synthesis of a vast number of empirical and theoretical studies on ecology, population dynamics, food production, production and effects of various types of pollution, technological innovations and their effects on the physical environment, and so on. The selection, organisation and interpretation of these studies subscribes to the same premise as that of *The Limits to Growth* which is common to neo-Malthusian theory namely a contradiction between physical finitude and exponential rates of growth of uncontained activity. Ophuls argues this contradiction applies to population dynamics though not in the proportions Malthus proposed in his 'dismal theorem in 1798', although in Ophuls's words 'the prospect for a species whose fertility continues to outrun its means of sustenance is still unrelievedly dismal'.[15] Populations still expand exponentially and so does demand for food but this exponentially growing demand for food cannot be indefinitely satisfied because of the ultimate limits of arable land, soil fertility and fresh water and the ecological limits of monocultural production, periodic crop losses and pollution deriving from agricultural production and consumption. This is so in spite of innovations in agricultural technology for increased productivity, for there are not only inherent physical limits to this expansion but there are costs incurred from higher energy requirements and lower ecological stability.

Ophuls makes an analogy between population growth and industrial expansion similar to that of *The Limits to Growth*. The expansion of industrial activity generates an exponentially increasing demand for minerals and fossil fuel which are resources finite in quantity just as the food supply is finite against an exponentially increasing demand owing to exponentially increasing sizes of populations. This is particularly grave in view of the rapid rate of industrial expansion which reduces the doubling time, i.e. the time in which the quantity of demand is doubled, and in view of the limited possibility of discovering and recovering new deposits. In addition, industry's 'consumption' of minerals and fuel, argues Ophuls, generates industrial pollution at an exponential rate of increase which in turn produces exponentially increasing costs of pollution control. These costs are ultimately self-defeating for they cannot succeed in containing pollution.

Ophuls examines the possibility of replacing fossil fuel power by nuclear power as a solution to depletion. In his assessment of the technical and environmental issues, this solution should not be retained for three reasons: (a) ultimately limited supply of a necessary nuclear-technology mineral, uranium, (b) the generation of heat by nuclear technology with concomitant climatically related ecological consequences, (c) more importantly, inestimable hazards of the negative effects of radiation on the health of eco-systems and human beings. He discounts fusion power and geothermal power as a solution for similar reasons, as well as the practical reason of the as yet undemonstrated feasibility of technology of fusion power and the limited possibility of geothermal power production not to mention the pollutant emissions which this technology produces. The solution Ophuls proposes on technical, economic, climatic and ecological grounds is thus solar energy.

In assessing these various forms of power technologies Ophuls examines the relation between restraint or reduction of levels of demand and continued availability of resources as a solution to the problems which each power technology generates with the exception of solar technology and suggests that neither conservation nor improved efficiency are immediate remedies. He bases his conclusion on the proposition that present institutions are incompatible with conservation. '*Continual* technological advance in energy conservation is not possible'.[16] Indeed, for Ophuls, present forms of technology, which he calls bulldozer technology, are inherently limited in regard to their actual and potential capacity to solve various problems related to ecological boundaries of the earth. He discounts, thus, the possibilities of a technological solution as such.

Ophuls defines ecological scarcity not as a simple measure of the physical constitution of the planet but as a relation between this physical makeup and the demands made upon it. His sense is a dynamic one as a relation of disproportion between supply as physically determined capacities and demands where the disproportion is ultimately grounded on the physical scarcity inherent in the earth's finitude and which ultimately determines the limits to demands as a measure of expansionary activity. It is thus a 'Malthusian scarcity of food (and). . .impending shortages of mineral and energy resources, biospheric or ecosystemic limitations on human activity, and limits to the human capacity to use technology to expand resource supplies ahead of exponentially increasing demands (or to bear the costs of doing so)'.[17] He proposes a sigmoid or logistic growth curve as representing the pattern of growth activity in relation to natural finitude such that the periods of acceleration and deceleration as well as equilibrium are ultimately determined by natural finitude. He suggests this is the growth curve of industrial civilisation and since this curve serves to define the present state of activity in relation to physical capability let us reproduce it here, in Figure 1.1.

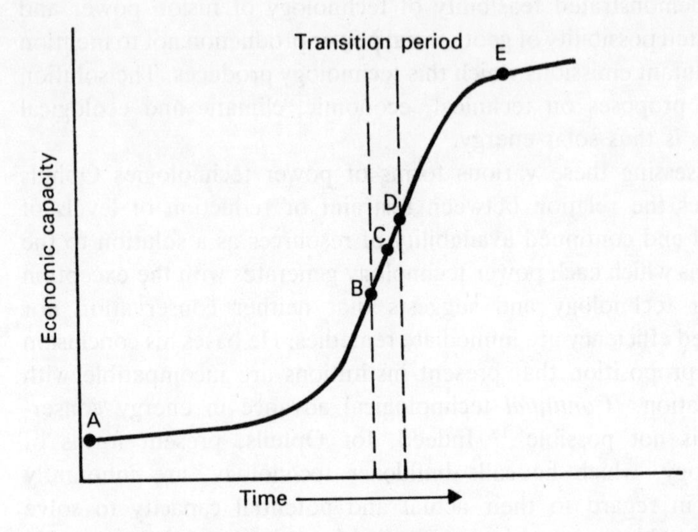

A steady state (beginning of accelerating growth);
B end of unrestrained growth (beginning of transition period);
C point of inflection (beginning of deceleration);
D end of transition period;
E terminal steady state.

Figure 1.1 Growth curve of industrial civilization

A to *B* represents activity of the last 300 years as accelerating growth where ecological and other resources necessary for growth are present in abundance. Following this, resources become scarce and at *C* deceleration begins. *B* to *D* represents the transition and *D* to *E* growth owing to momentum but, he adds, 'the ecological abundance that fueled accelerating growth begins to disappear, and the first warning signs of ecological scarcity are quickly succeeded by various negative feedback pressures that start to choke off further growth'.[18] Deceleration continues until equilibrium *E* is attained.

The zone of transition is critical since the changeover from accelerating to decelerating growth occurs in a very brief time and since this is the period during which the limits of finitude of ecological scarcity become evident. Ophuls claims that we are presently in this period, or past the point of inflection C. There is thus an urgency to modify the pattern of economic activity in line with the limits of ecological scarcity. This conclusion echoes that of *The Limits to Growth* and concurs generally with those of other neo-Malthusian studies of trends and projections including subsequent reports to the Club of Rome. Indeed, the general conclusion of all neo-Malthusian assessments of present patterns of activity including the activity of breeding is the necessity of restraint to an extent determined by ecological exigencies. This proposal is the characteristic feature of neo-Malthusian social and political theory.

NEO-MALTHUSIAN SOCIAL AND POLITICAL THEORY

The social and political theory of neo-Malthusianism rests on the premise that human wants are infinite but the physical substructure necessary for their satisfaction is finite. This posited contradiction is resolved by an activity consistent with natural exigencies and physical finitude. This is essentially an activity of restraint both in the quantitative sense of limiting wants and in the sense of limiting the technical means of satisfying them to those compatible with ecological exigencies. The activity of restraint is politically determined such that allowable forms and degrees of activity are specified and monitored. It is thus an externally imposed restraint on the individual's activity. The individual does not freely define objectives and does not freely determine means to achieve them, but is rather subordinate to the political structure as one of control and surveillance of activity in line with ecologically determined criteria.

In examining this theory we shall argue that the premise of unlimited human wants is the Hobbesian liberal assumption of human nature as insatiably acquisitive. This classical liberal anthropology has been interpreted[19] as being bourgeois in the sense of conceiving human nature on the model of the private, bourgeois individual who achieves his or her essence in the activity of private accumulation, or, in other words, private property. Neo-Malthusianism, we shall argue, implicitly assumes a human nature and the form of human nature it posits is private. Implicit in neo-Malthusianism is that the activity of private appropriation is basic to human nature. We shall argue that this anthropological assumption of private nature together with the assumption of insurmountable physical finitude underlie the theory of politically determined restraint. Physical ecology, rather than determining the form of this activity, merely sets the quantitative boundaries within which activity can proceed. Since human nature is essentially private in neo-Malthusianism, this activity as an expression of human essence is private appropriating activity. The political theory of restraint thus is a theory of quantitative restraint of the amount of private property an individual may acquire.

Our interpretive hypothesis that a Hobbesian liberal anthropology underlies neo-Malthusian theory seems at first sight incorrect. Classical liberalism, after all, is a champion of individualism and neo-Malthusianism is vehemently critical of individualism. We find in neo-Malthusianism a depiction which is not altogether flattering of the independent individual as being excessive, irresponsible, chaotic, defying rather than obeying the laws of nature and creating harmful and destructive ecological and social conditions. He or she lacks a sense of moderation and control and depicts the predominant role as that of maximising a personal share of resources and of extracting as much from nature as possible. The individual exercises no restraint on personal appetites and, lacking all sense of frugality, remains insensitive to nature's limited capabilities. Ecological damage is thereby inflicted in proportion to the excesses of uncontained appetites.

This depiction of the individual stands in sharp contrast to the post-Hobbesian classical liberal conception, particularly that of Adam Smith. For neo-Malthusians the unrestrained individual leads to ecological and social destruction and for classical liberalism this same unrestrained individual leads to social well-being or in Adam Smith's words, 'the greatest good for the greatest number'. The difference lies in the opposing judgement of the same perception of the individual and this common conception derives from the common anthropological theory of human nature. For both classical liberalism including Hobbesian liberalism and neo-Malthusianism, unrestrained activity is an expression of human nature. But this free activity of classical liberalism is conceived as being a rational activity. It is rational because it is determined by reason and it is the individual who is the sole source of rationality. He or she defines personal goals, and develops devices to achieve them. Rational unrestrained activity is not the activity of Hobbes's state of nature 'of war of all against all', although that which is exercised in such a natural state is indeed unrestrained. That which distinguishes between rational unrestrained activity and irrational unrestrained activity is the relation between individuals. Since reason is the property of the individual no external source of authority should tamper with it. Unrestraint, then, refers to the freedom of the individual to exercise his reason according to his own volition. This, of course, implies a relation of equality in the exercise of reason such that one does not constrain or impede the exercise of reason of the other. Competition is conceived as a wholly rational means by which individuals seek to achieve their own goals. It is a modality of action of equal and free individuals. It is in this exercise of individual rationality that human nature is expressed freely.

The key to the opposing judgement of individualism between classical liberalism and neo-Malthusianism is the relationship drawn between private individual rationality and public common good. For classical liberalism there is consistency, compatibility between the two. The interest of the private individual and the common interest is dependent on the private interest in the sense that unhampered, unimpeded activity to achieve private interests satisfies as well the common interest. There is no contradiction between the two since the common interest collapses into the private interest and becomes one with it. In neo-Malthusianism the relation between private and public interest is not one of harmony but of contradiction. Unrestrained activity in the pursuit of private interests leads to what Garrett Hardin calls 'the tragedy of the commons', i.e. depletion of resources, uncontrollable pollution and overpopulation. The resolution of this contradiction is evident for neo-Malthusianism. It is the subjection of private interests to the common interest.

The neo-Malthusian notion of unrestrained activity is not entirely equivalent to that of classical liberalism although the differences are minor and do not affect but rather serve to reveal their common anthropological basis. The 'unrestraint' character of the activity is conceived as relating more to the quantitative aspect of this activity than to the form of the activity itself. Private reason gets 'carried away', in a sense, and makes quantity its objective. Producing or acquiring more and more of this and more and more of that is unrestrained activity. It is propelled by unlimited wants including the want to have children. But although this quantitative aspect of unrestrained activity is emphasized, it is linked to the free individuals of classical liberalism (and indeed is enlarged to include the free corporate groups of neo-liberal pluralism) by the postulate of human nature as infinitely acquisite.

If human wants are infinite they will be converted into quantitatively maximising activity by the free expression of the individual (or group). The individual or corporate group cannot be entrusted with the freedom to define private goals and proceed to achieve them unhampered by any external authority. The quantitative emphasis of unrestraint thus does not contradict the classical liberal unrestraint as unimpeded activity of the free subject. But the implications for social and political theory are dramatically different.

For classical liberalism this activity is rational since the individual is the source of reason. The public authority, the state, has no other function or *raison d'être* than to secure and maintain peace and order such that the free individuals can comply with their reason. For neo-Malthusianism this activity is irrational as the reason of the free individual is the reason of private acquisition. It is limited to calculations promoting the activity of private maximisation. In neo-Malthusianism the standard of right, rational conduct is not the free individual. The reference point of such conduct is external to the reason of private acquisition; it is physical nature in its finitude. It is the finitude of physical nature against which activity must be measured which determines ultimately the rationality or irrationality of activity. Moderate, limited activity is rational; unlimited, uncontained activity is irrational.

This comparative depiction of classical liberalism and neo-Malthusianism is, of course, very sketchy and very general but this very generality helps to identify their basic similarities and differences and to suggest a theoretical continuity between them deriving from their common anthropological premise of human nature. In this comparative sketch is a general statement of neo-Malthusianism which points to the theoretical agreement between neo-Malthusian thinkers. If unrestrained activity in a world of physical finitude has the effect of economic and social crisis of overpopulation, pollution and depletion of resources, then such activity must be modified to conform with the physical and ecological capabilities. This general statement, however, does not imply that there is one monolithic neo-Malthusian voice as to the passage from exponential growth to a contained growth including, as some suggest, zero-growth. It does not suggest that there is no variation in the political theory of contained activity. A 'coercive theory' is advocated by Garrett Hardin and William Ophuls, a 'market theory' by Kenneth Boulding, a 'market/moral theory' by Herman Daly, a 'naturalist-necessity theory' by Ratko Milsavljevic, a 'global integration' theory by Aurelio Peccei, Lester Brown and Harold and Margaret Sprout all of whom propose in a more or less explicit way means of achieving a degree of activity sufficient and necessary for the reproduction of a system as a 'steady state' or a state of equilibrium as determined largely by ecological exigencies or physical necessity.

THE THEORY OF COERCION

'A mutual coercion, mutually agreed upon by the majority of the people affected' is the definition of coercion Garrett Hardin attaches to his political theory. His influential 'The Tragedy of the Commons'[20] has the implicit objective of determining a basis for judging activity and following this the more explicit objective of specifying means for controlling activity such that a certain form and degree of action be positively and negatively sanctioned, i.e., politically controlled. The implicit argument in Hardin's thesis of the necessity of political control, based on J. Fletcher's *Situation Ethics*,[21] is that an activity can be judged in all aspects including ethical and moral, in relation to the total environmental context in which it is performed and more specifically to its effects. If the effects are injurious to others or to the general environment and ecology, the activity is judged as being wrong even immoral and such activity must enter the realm of public surveillance and control.

In developing this thesis Hardin selects the phenomenon of population expansion and links it to the phenomenon of growing pollution as a cause/effect relation or, as he puts it, 'the pollution problem is a consequence of population'.[22] High demographic density generates high degrees of pollution injurious to society and ecology. The Hardin thesis proposes a more basic cause to the expanding population and growing pollution, namely the freedom to breed. As the rate of birth increases and rate of death decreases owing to longer lives, there is more and more activity weighing more and more heavily on the resources of the planet including the common air and common water which constitutes the tragedy of the commons. Formerly, activities determined freely by individual wants had limited effects on society and ecology but as activities have multiplied the effects of these same activities stretch across the whole ecological and social or public system. Hence, a public instance must intervene to limit 'freedoms'. The main function of the state, then becomes the function of legislating temperance and creating a system of administrative law for judging temperance or moderation of action in specific circumstances and penalising the excesses. 'Since it is practically impossible to spell out all the conditions under which it is safe to burn trash in the back yard or to run an automobile without smog-control, by law we delegate the details to bureaus'.[23] A state mechanism is thus put in place for containing activity, including and especially the activity of procreating, by threat of sanctions since 'it is a mistake to think that we can control the breeding of mankind in the long run by an appeal to conscience'.[24]

William Ophuls takes up Hardin's coercion theory and builds on it retaining the anthropological assumption of human activity as acquisitive activity propelled by unlimited wants and the assumption of scarcity as determined by physical finitude. Ophuls likens this theory to Hobbes's *Leviathan* both on the level of assumptions and on the level of political authority. The similarity is drawn on largely superficial grounds but it is not mistaken. On the contrary, it draws attention to the more profound agreement between classical liberalism and neo-Malthusianism.

In Ophuls's reading, Leviathan is a public authority which stands above individuals in order to preserve peace and order for all. Individuals have a need of such a sovereign for it is this which distinguishes between civil society and state of nature where individuals naturally compete indiscriminately for scarce resources and where might prevails over right. In Ophuls's view, thus, the condition of scarcity and unlimited human wants led Hobbes to propose a Leviathan or sovereign with power to constrain individual action. This Leviathan has the enlightened consent of individuals for reason dictates that to be rationally constrained is better than to perish.

It is this reading of Hobbes which Ophuls relates to Hardin and which figures predominantly in the model of his theory of coercion. This theory is also an adaptation of the Rousseanian notion of 'general will' along the same lines as this particular adaptation of Hobbes's *Leviathan*. Ophuls retains from both notions a sense of a public instance or public policy which is superior to individual wills or interests and stands above them defining form and degree of permissible activity and imposing this definition on individual wills and individual activity. For Ophuls, as for Hardin, this is an external imposition hence coercion on individual will and action. It is so conceived because normal individual activity is conceived as being contradictory to the common good. The individual sees only immediate personal interests and as these are the reflection of unlimited wants, will naturally and spontaneously behave in ways incompatible with or injurious to the common good. Individuals therefore cannot be left to their own devices but rather require other means in order to comply with the common good. These 'other means' are means they would not normally arrive at alone as their immediate interests prevent it. It is in this sense that public authority is external and in this sense coercive. But the general will nevertheless has the ultimate consent of the individual wills for individuals do possess a reason recognizing the necessity of the general will to preserve a climate of security where activity is contained in a way compatible with social and ecological well-being.

The form which Ophuls gives to the general will is a steady state political economy which is a reduction of activity to that level compatible with ecological exigencies. The form of activity here is not qualitatively changed, it is merely quantitatively limited. An economic private entrepreneurial activity is limited as to the quantity of mineral resources and fossil fuel is may be permitted to produce for marketing or for use in manufacture, and in the amount of toxic wastes it is permitted to discard into the air, water or soil. Technological activity is reduced as well and here the reduction may involve the limitation or even abandonment of some forms of technology, for example, nuclear in favour of solar energy. Parents are limited in the number of children they are permitted to have.

Ophuls proposes in passing the notion of 'ecological contract' formally comparable to the social contract of classical liberalism which suggests the sense of determination of his general will and in turn of the steady state society or steady state political economy. It is a political pact between people defining not only social, political and economic relations between themselves as is the form of the social contract, but rather relations between themselves and physical nature.

> Ecology and politics are now inseparable; out of prudent self-restraint, if for no other reason, a valid political theory of the steady state will be obliged to give the same weight to ecological harmony as to social harmony. Thus, just as it was the task of the seventeenth- and eighteenth-century political philosophers to create the social-contract theory of government to take account of the new socioeconomic conditions and justify the political ascent of the bourgeois class, so it will be the duty of next generation of philosophers to create an 'ecological contract' theory

promoting harmony not just between men, but also between man and nature.[25]

As opposed to the theory of social contract which conceived of social harmony as an effect of equality and individual freedom, the implication of Ophuls's theory of ecological contract is that a social and ecological harmony will be the effect of limited activity. In the former, discord and conflict emerge from infringements on individual freedom where some 'do unto others as they would not want others to do unto them' and create an inequality by might. The *raison d'être* of Hobbes's *Leviathan* is to oversee activity and bring it into line with equal as opposed to indiscriminate competitiveness. This is the harmony of social, political and economic relations because it is in the interest of all. Of course, the interest of all here is the private interest of the entrepreneur. It is the interest of the class of entrepreneurs or capitalists to secure a political framework in which rules of competition are those of the free market, i.e. free circulating capital. This thesis has been explored sufficiently elsewhere[26] to warrant simply a reminder of it here.

In Ophuls's embryonic theory of ecological contract the implication is that discord and conflict – to which could be added misery, poverty and organic decay – are the effects of another kind of indiscriminate activity. It is excessive activity which pollutes and depletes nature and makes for unemployment and poverty at the social level as a result of population size. It is in the interest of all therefore to limit this activity. In this way, political constraints on the 'macro-level' convert to 'micro-freedom' at the individual level. The individual may still pursue his private interest as long as he does so in ecologically non-injurious ways. 'Nor need the right to own and enjoy. . . property be taken away, only the right to use private property in ecologically destructive ways would be checked'.[27] In this context the sense of Ophuls's political theory of coercion is captured best by this passage:

> the socioeconomic machinery needed to enforce a steady-state political economy need not involve dictatorial control over our everyday lives, it will indeed encroach upon our freedom of action, *for any social device that is effective as a hedge will necessarily prevent us from doing things we are now free to do or make us do things we now prefer not to do.*[28] (author's emphasis)

It is not entirely coincidental that Hobbes recurs in this critical theory of growth and political theory of steady-state society. The theoretical similarities are greater than those made explicit in the theory. We suggested earlier that they both share the same anthropological assumption of human nature as unlimited wants propelling an acquisitive activity. This specific premise is posited as a universal principle – indeed an extrasocial principle – and as one capturing the sense of life and existence. It posits human nature as existing prior to social relations and thus as underlying all forms of social relations. We suggested that this premise in its form and in its very claim to universality is rather a socially determined premise and specifically an entrepreneurial class-determined premise. The critique of growth is mediated from a perspective which posits private entrepreneurial activity as natural and unlimited even in its effects of crisis of society and ecology. It is this perspective which mediates the quantitative solution of crises for if this activity is unlimited as propelled by infinite wants in a physical world of finitude, it is not unlimitable. Reducing it is the means of retaining its form and content. It is the means of retaining relations of private property.

THE MARKET THEORY OF RESTRAINT

We find similar reasoning in Kenneth Boulding's market theory of the steady-state and in its adaptation by Herman Daly. Kenneth Boulding's[29] theory of the steady-state adopts the premise of physical finitude as a closed planetary system of 'spaceship earth' into which he incorporates human activity. From this follows his notion of interrelatedness or interconnection between inputs and outputs constitutive of the structure of the closed system, the three critical inputs being matter, energy and information. Social activity, in other words, is conceived formally on the model of physical matter and energy, and it is this model which provides quantitative indications for containment of activity. Physical finitude and thermodynamic laws of nature become the determinants of human activity.

To define more specifically the levels of containment of activity, Boulding distinguishes between two conceptions of the planet as a physical substructure or environment of an economy. One conception is of a planet as an open system which takes raw materials from the external environment as inputs, and ejects effluents as outputs to this same external environment. The other conception is of a closed system with no unlimited, external environment and where reservoirs of raw materials and effluent absorption are limited and tied to the internal operations of the system. His first principle of physical finitude makes this latter notion of economic activities on the planet a conceptual and practical necessity. He proposes, thus, the notion of 'stock maintenance' for a closed system in place of growth of an open system. The 'stocks' are population and capital and their level of maintenance is determined by the closed system's limited capabilities. Stock maintenance, accordingly, is the critical factor of the steady-state. Breeding activity and economic activity is limited to the level of stock maintenance.

In Boulding's conceptualisation of the steady-state, a political question is introduced only at the level of regulation of activity consistent with stock maintenance levels, since the whole matter of physical finitude, interrelatedness, holism and equilibrium are determined by organic laws and the recognition of these laws by reason makes limited activity a necessity. How to limit the birth rate and economic activity, both production and consumption, and how to keep this activity contained is the political question. The answer is Boulding's political theory. It consists of a market theory of regulation of activity where, in the case of population stock maintenance, for example, a system of licensing to have children would operate such that a number of licences would correspond to replacement fertility, and be bought and sold on the open market with financial compensation for unused licences.

Herman Daly proposes a theory of steady-state very similar to that of Boulding, whom he indeed acknowledges as a mentor. He proposes it as a new paradigm for economic thought to replace the currently dominant neo-classical paradigm which he calls the paradigm of 'growthmania'. His paradigm comprises a more comprehensive statement of social, political as well as economic activity for which he draws from John Stuart Mill's *Principles of Political Economy* as well as Kenneth Boulding's theory of spaceship economy. It is one which posits physical finitude and the ecological principles of interrelatedness, holism and equilibrium as first and determining principles of a social and economic system. It explicitly rejects the assumption of infinite wants but retains an assumption of private property as an expression of individual freedom. We suggest that the political theory of the steady-state, which Daly contends is neither capitalist nor socialist but rather 'distributive', attempts to reconcile limited or controlled individual activity with private property. Although in many respects this theory differs from that of Hardin and Ophuls it agrees with these latter in this fundamental sense.

Daly defines steady-state as 'an economy in which the total population and the total stock of physical wealth are maintained constant at some desired levels by a "minimum" rate of maintenance throughput (i.e. by birth and death rates that are equal at the lowest feasible level, and by physical production and consumption rates that are equal at the lowest feasible level)'.[30] He justifies this steady-state as physical necessity in terms of his first principles. The main problem as Daly sees it, then, is the working out of the means and mechanisms of achieving a steady state and of maintaining it.

What is of interest to us here is not so much the basis of definition of optimal and minimal activity which is obviously knowledge of physical and ecological processes but rather of the means of maintaining the activity within allowable limits. Here Daly proposes a notion of social institutions of control which he proposes 'are of three kinds: those for maintaining a constant population, those for maintaining a constant stock of physical wealth, and those governing distribution',[31] such that maximum control is reconciled with individual freedom. 'In all cases the guiding design principle for social institution is to provide the necessary control with a minimum sacrifice of personal freedom, to provide macrostability while allowing for microvariability, to combine the macrostatic with the microdynamic'.[32]

A detailed system of controlled and limited licensing to have children equally distributed amongst the population with possibility of transfers and compensation of unused licences, is proposed as a means of maintaining steady stock of population in the steady state. A detailed system of market-regulated quotas controlling the use of natural resources and production is proposed as a means to govern economic activity. That is, the quotas are issued by the state but are freely transferable by sales or gift and the market principle operates in the regulation of prices of quotas in relation to the aggregate rates of depletion of material resources and aggregates of rates of pollution. Thirdly is proposed a system for controlling the distribution of wealth involving its confiscation above a fixed upper level and redistribution guarantee of a lower limit in line with natural finitude and ecological first principles. This is not, in Daly's reasoning, a violation or attack on private property but rather, citing approvingly John Stuart Mill, a defence of it. 'Private property is legitimated as a bastion against exploitation. But this is true only if everyone owns some minimum amount. Otherwise, when some own a great deal of it and others have very little, private property becomes an instrument of exploitation rather than a guarantee against it'.[33] Activity along these lines is, in Daly's as well as in Boulding's terms, moral activity. It is activity limited in quantity but identical in form with that activity posited as essential to the expression of human nature by classical liberalism; it is the activity of private appropriation.

NATURALIST-NECESSITY THEORY

Ratko Milisavljevic proposes what I shall call a 'naturalist-necessity' theory in his *Environment, idéologie et science*.[34] The neo-Malthusian essential argument is the prominent feature of this work namely that expanding population and capital growth is hitting against the ultimate limits of physical nature and the only way to avert an ecological collapse is to bring activity in line with the limits of physical finitude. The interesting feature of this work is the conception of science and ideology in regard to activity. Both science and ideology are conceived as being conceptual structures of reality mediating social activity. The difference between the two, however, is the critical factor for the political theory implicit here. Science is conceived as being the recognition of natural laws and thus scientific theory is that which depicts faithfully and accurately organic causal processes and natural laws. As he puts it, 'la pensée scientifique est le reflet psychologique de la réalité naturelle' (scientific thought is the psychological reflection of a natural reality).[35] On the other hand, ideology of whatever form is conceived as being a distorted representation of reality and that this distortion is mediated by private interests of dominant ruling groups throughout social history. The present growth ideology of both capitalist and state socialist systems dictates increased production and consumption but leads, according to Milisavljevic, to 'l'épuisement de l'énergie et des matières premières, la pollution de l'écosphère, la surpopulation' (exhaustion of energy and of natural resources, pollution of the ecosphere and overpopulation).[36]

The solution is obvious to Milisavljevic. It consists of replacing ideology by science as it is the latter which can specify the limits to activity as determined by organic exigencies and so lead the way to the creation of a single planetary society in unity with nature. In this solution is contained Milisavljevic's implicit political theory although in his terms science is value-free and apolitical, hence he would term his theory scientific as opposed to political.

> Although political means have prevailed in the course of history, they are not capable of resolving the relations between society and nature for they are means of generating and reinforcing social power. What is required now is the exact opposite: no longer should there be social power except the power inherent in the determinism of natural and social phenomena. . . This model of new organisation of society is not a matter of human thought but something given in the natural conditions of existence, human society being willing to conform to the means of science and knowledge.[37]

Apart from the question of replacement of ideology by science which is a political question, there is a political notion here of activity as regulatory. It is actively determined in form and degree by physical necessity. The recognition of this physical necessity is the scientific mediation between nature and human activity. This becomes good, moral activity as activity harmonising the social and the natural. 'C'est dans la pensée scientifique que l'on a reconnu les forces du bien et dans l'idéologie celle du mal (In scientific thought, one recognises the forces of good, in ideology the forces of evil).'[38] This is a proposal for a science-led global integration for the creation of a planetary society.

GLOBAL INTEGRATION THEORY

A proposal for global integration is also highlighted in a number of integration theories namely those of Aurelio Peccei, Lester Brown and Harold and Margaret Sprout. These integration theorists propose global or planetary economic and political integration as a solution to the neo-Malthusian problem of unlimited growth. It is what we may call a theory of rationalised activity. The underlying notion of this theory is efficiency, consuming the minimum necessary energy both at the level of individual activity and at the level of total global activity. It is thus a theory of combining or integrating conservation-efficient individual activities in a conservation-efficient total collective or global activity. Rationalisation is the means proposed for quantitatively reducing the demands made on the planet and bringing them more in line with the finite capabilities of the planet in terms of both natural resources and pollution absorption. It poses the ecological question as a problem of waste and overconsumption resolvable by elimination of duplication and of excesses. Global integration becomes the key, then, to rationalisation.

An economically and politically integrated global society is proposed as best suited to satisfy the naturalist-determined necessity of rationality. Such a global society replaces the nation state which on the basis of the principle of rationalisation is inconsistent with natural finitude. Militarism, duplication of material infrastructure, economic nationalism as features of national sovereignty in a plurality of nation-states are antithetical to rationalisation principles. The rationalist sense of the neo-Malthusian integrationist critique of the nation state is captured in this passage by Lester Brown in *World Without Borders*.

> Measured in social terms, the cost of maintaining the existing system of independent nation-states is extremely high. It is largely responsible for spending more than $200 billion worth of the world's public resources for military purposes, and for the artificially high consumer prices for goods which are inefficiently produced behind the protection of national tariff walls. It is also responsible for a great deal of redundancy in scientific research.[39]

Harold and Margaret Sprout reiterate the same argument.

> Neither the United States, the Soviet Union, nor any other national polity can sustain the costs of global hegemony, an endless arms race, or a prodigal array of fantastically expensive prestige projects without progressively weakening the domestic society from which all power and influence are derived; and. . .that very few, if any, of the underdeveloped societies can meet the costs of modernization while simultaneously caught up in the vicious race between population increase and economic productivity, especially when accompanied by expensive adventures in militarism and international politics.[40]

Aurelio Peccei emphasizes amongst other things the cost of duplication of material infrastructure in the present geo-political pattern of nation-states.

> The excessive growth of military forces, extorting six percent of the product of human labor every year, is not the only support

on which this senseless partitioning of the world can rely. In addition there is a proliferating diplomatic network and a burgeoning array of information or propaganda services. Their usefulness is certainly in question at a time when radio, television, and the press put the news within everybody's grasp; when the telephone, telegraph, telex, and airlines span the world; when scores of satellites are constantly on station over the Earth; and when journalists disclose everything that is even remotely of interest. These superstructures can titillate the taste of a public that loves military display, fanfares, parades, regalia, secret funds, and cloak-and-dagger novels to such a degree that they forget how superfluous and costly all this is.[41]

The form of a world state however is not worked out nor is there a specific proposal for the form of its regulating activity. It is, in principle, a structure for co-ordination, planning and regulation of activity to degrees or extent compatible with ecological principles and natural finitude. The features which integrationists attach to a world state are simply those which they see the nation-state as lacking. In support of the proposal for a world state with power to coordinate and regulate activity, Peccei argues,

> In its present form, *the world is both structurally and politically ungovernable*. In the current state of affairs, there is absolutly no possiblility of instituting a 'New International Economic Order'. . . . The essential problem to resolve is that of incompatibility between the existence of the inward-looking, sovereign nation-state and the well-ordered management of human affairs on a global scale, which has become a fundamental imperative of our times.[42]

These features appear to be no other than those of the present nation-state transposed on to a single world state. Political sovereignty is transferred from the nation state to the world state.

From the rationalisation principle of integrationist theory follows a generally favourable assessment of the multinational corporation as an organism of economic global integration. Lester Brown argues that

> In its efforts to achieve the most efficient possible combination of productive resources, the MNC contributes to the creation of a more equitable world order. It raises investment capital in countries where it is abundant and interest rates are low, investing

> in poor countries where interest rates are high. Likewise, it attempts to locate its more labour-intensive operations where wages are lowest, thereby helping to raise incomes in poor countries. The MNC uses the most efficient technology available irrespective of the country of origin. The net effect world-wide is to provide a higher level of living for a given use of resources and effort than would be possible without MNCs.[43]

The assessment of the pollution production record of the multinational corporation and of its consumption of natural resources is unfavourable on the grounds of excessiveness. It is the conception of economic activity which casts light on the sense of political regulation of the world state in integrationist theory. It is a regulation of limiting activity quantitatively to ecologically tolerable levels. It is this common conclusion of the variety of neo-Malthusian political theories which unites them and which reveals something of their common ideological form.

To identify the underlying premise in neo-Malthusian theories as Hobbesian anthropology and to demonstrate the coherence which this premise brings to the various elements of neo-Malthusian thought is not to account for this anthropology, itself. That is, it does not answer the critical question of how an assumption of human nature as 'possessive individualist' becomes the cornerstone of the conceptual structure. It is precisely this question which C. B. Macpherson explores in his analysis of the possessive individualism of the classical liberal thought of Hobbes and Locke.[44] It is a question which leads him to competitive market relations or relations of production of competitive capitalism as containing the key to the possessive individualist assumption of human nature. It is an assumption which conveys the nature of the bourgeois individual not universal human nature as it claims.

Marx's analysis of the capitalist mode of production in *Capital*[45] demonstrates the actual social constitution of the bourgeois individual and of liberal individual assumptions, notions and values which while being specific to this individual are posed as being universal. His analysis points to the level of circulation of capital as determinant of liberal ideological notions but this determination is itself the effect of the whole productive structure. It is relevant to a discussion of possessive individualist anthropology and its basis of determination to recall however briefly some of these considerations. One is the feature of the capitalist mode of production as comprising two distinct 'separate' spheres – the sphere of circulation and the sphere of production. This separation of exchange from production is the effect of the commodity of labour-power which owes to the separation of the direct producers from the means of production as the distinct structural characteristics of the capitalist mode of production. This separation captures the structure of the commodity including labour-power as separation between use-value as the concrete form which designates its utility and exchange value, the abstract form of value indifferent to the concrete content of commodity. As regards the commodity, unique to this mode of production, namely labour-power, its exchange-value becomes confined to the sphere of circulation and its use-value to the sphere of production at which level labour-power produces an excess of value for that for which it was exchanged. But this excess (surplus-value) does not appear at the level of exchange, for indeed that which is exchanged is labour-power not labour-power-in-use, where all commodity owners (including labourers as owners of labour-power) confront each other as equal exchangers. They are also free to enter into contracts of their choice, the sole criterion being the maximisation of private property. The buyer of labour-power chooses the cheapest offer and the seller of labour-power chooses the highest bidding buyer in a free flow of the labour-power commodity. Notions of individual equality and individual freedom emerge, hence, as ideological effects of this structure of productive relations. The exchanger sees himself as master of his fate, free to pursue his objectives as he himself defines them and free to develop or choose means he deems appropriate to achieve them.

This mode of equality and this mode of freedom, however, are not abstract nor universal but rather specific to the activity of private appropriation and, what is more important, they are relevant to and indeed determined by the capitalist relations of production. One insight which Marx's analysis provides is that these notions are ideological in the mystification sense as this equality is only an equality of exchangers at the level of exchange not of producers in relation to proprietors at the level of production, the level at which value is produced. Another insight is that this structure rather than being one of equality is one of class dominance. It is a dominance of the class of proprietors.

If the very notions of possessive individualism which are the effect of this structure mediate an understanding of these relations they can only give rise to a bourgeois class understanding of these relations. Neo-Malthusian thought is precisely such an understanding and the proposals for change are marked by the limits of such an understanding.

2 Expansionism

If our interpretation of neo-Malthusianism is correct, it is not theoretically adverse to its most antagonistic opponent, namely expansionism. Its roots in liberal Hobbesian anthropology bring it into close theoretical proximity to expansionism as it, too, is essentially a liberal Hobbesian theory of human nature and society. Expansionism as expounded by R. Buckminster Fuller, John Maddox, Wilfred Beckerman, and Herman Kahn is not a current of thought normally associated with ecological theory except as a vehement critique of ecological theory in general and neo-Malthusianism in particular. The themes of ecology, environment, natural resources figure very prominently in the writings of these authors although not from a sense of urgency to save them from imminent collapse as is the case of neo-Malthusianism. The basic general argument is that the present form of economic expansion poses no threat to the viability of the environment but, on the contrary, contains the key to the solution of any environmentally related problems however and whenever they may arise.

To ecologists of any perspective this appears as a very cavalier, ecologically insensitive position. It is merely a means to conceal the injurious effects of expansionary activity on the environment and therefore one task of ecologists must be to unmask these effects in theory by revealing the negative effects of expansionary economic practice. Neo-Malthusianism adopts this stance as do eco-developers not to mention radical ecologists as we shall see later. Expansionary theory is considered thus to be not a form of ecological theory but rather anti-ecological theory. A consideration of expansionism as a theory of ecology may therefore offend the most passive, dispassionate ecologist. Our decision to include expansionism in the context of ecological theories, however, is prompted not so much by an interest in expansionism as just one form of ecologically sensitive or insensitive social theory, but rather by what its common theoretical underpinnings with neo-Malthusianism imply for the thrust of neo-Malthusian critique. In examining expansionism, we are, in a sense, inserting a parenthesis to qualify aspects of neo-Malthusianism rather than presenting another critical form of ecological theory.

The expansionary thesis can be summarised as a totally optimistic statement regarding the present social condition, the earth's capabilities and resources and future prospects for expansion. The roots of this thesis can be traced to Hobbesian liberalism similar to those of neo-Malthusianism with some significant differences which yield apparently opposing conclusions. There are three interrelated premises to this thesis, or more precisely, three interrelated aspects of one basic premise, namely a Hobbesian anthropological tenet of human nature which contains a particular notion of human subject, of the world external to the subject, and of the subject's relation to the external world.

The central notion of this anthropological tenet is the subject to whom is ascribed an inner materialist essence conceived as an unending quest for satisfaction of infinite appetites. Hobbesian man, of course, is no mere passive hedonist as this formulation would imply. The anthropological tenet, rather, vests the individual with the properties of subject, as having determining powers both as regards the appetites and the means of satisfying them. The subject imparts form on the appetites by positing them and satisfying them. They acquire materiality in the unified process of positing and realising. These infinite appetites which thus inhere in the human being are always mediated by reason and it is this which confers on the human being his quality of subject. When expansionist Herman Kahn depicts the essential nature of man as being materialistic, restless, probing and accumulating, he is taking up Hobbesian man as an insatiable being whose reason defines wishes, hopes and desires and finds ways of satisfying them such that new ones are generated in an infinite forward movement. He tells us that 'people have almost everywhere become curious, future oriented and dissatisfied with their conditions. They want more material goods and covet higher status and greater control of nature'.[1]

This same anthropological tenet posits nature as external to the subject or as the subject's 'other'. It is conceived as being passive matter, void of any value in itself. Any meaning or value is imparted to it by the subject in purposive activity. Hence nature as external world is conceived in relation to the subject or more accurately in terms of the subject's relation to it, as external world has a sense only for the subject. Since in this anthropology human nature is conceived as being a quest for satisfaction of appetites rather than satisfaction *per se*, the formative effect of this activity on the external world or nature is conceived as being equally without limits or bounds. Accordingly, the world grows and expands endlessly relative to the limitless formative activity of the subject.

The basic Hobbesian anthropological tenet in expansionism contains no contradiction between subject and external world but rather a wholly harmonious relation between them. Any limitation inhering in either the subject or the external world is a totally alien notion. The subject is driven by an inner essence to grow, and expresses this essence in a world which offers no insurmountable constraints or qualitative limitation but rather defies any boundaries.

This same Hobbesian anthropological tenet in neo-Malthusianism contains a contradiction between subject and external world. On the one hand, human nature is infinitely acquisitive but on the other hand the world in which the subject seeks to satisfy his appetites is naturally finite. The world cannot provide the infinite raw material which the subject requires for his essential activity. The subject cannot express his essence in the absence of these raw materials. It is a contradiction which, we saw, is resolved in neo-Malthusianism by the imposition of external constraints on the subject in the form of political coercion. This externally imposed restraint is an ecologically dictated requirement.

Not only is there an absence of such a contradiction in expansionism, but nature is totally deprived of any value in itself. It acquires sense in and through the subject's activity. The world, in other words, is the subject's world. This is the sense of Miguel Ozorio de Almeida's picturesque formulation of a basic expanionist assumption that the 'environment is supposed to serve human rather than reptilian, ruminant, pachydermic, or any other interests.'[2]

The subject's world is not conceived in Hobbes as being limited to physical nature. It is both physical nature and society collapsed into one undifferentiated entity of organic causality which submits to the subject's mastery and control. Indeed, in this perspective, the relation between the subject and the external world is a purposive rational relation. That is, the subject's formative activity is essentially a technical activity in which the subject seeks to satisfy his insatiable appetite of acquisition by mastering organic processes and combining correct cause/effect relations.

The subject here is dependent on the contingency of the external world and its organic causal natural laws, a knowledge of which extends his freedom – a freedom to manipulate in a way to control an outcome. Indeed, society in this Hobbesian liberal conception is an extension of nature in which the same natural laws operate. Consequently, the activity of Hobbesian man toward other men retains the same strategic, instrumental form of creating conditions in which the behaviour of others can contribute to the maximisation of his chances of achieving his own goals. At any given moment the degree of technical mastery of the external world – nature – is a measure of his essential development.

The Hobbesian notion of essential human development as technical mastery is the notion of progress in expansionism. We see this notion of progress in Herman Kahn's characterisation of technical innovation, a process of extending mastery over nature and one which is absolutely limitless.

> capacities for and commitment to economic development and control over our external and internal environment and concomitant systematic, technological innovation, application, and diffusion, of these capacities are increasing, seemingly without foreseeable limit.[3]

In Kahn's characterisation of history as comprising two Landmarks, namely the agricultural and industrial revolutions,[4] the notion of technically mediated progress is highlighted.

EXPANSIONIST CRITIQUE OF NEO-MALTHUSIANISM

These three notions, subject, external world, subject's relation to his world, are the key to the optimism of the expansionist thesis. Driven by his infinite, insatiable appetites, the subject continuously expands his technical mastery over the world extending his freedom and expanding his world. Subject, technical activity and nature are in harmony. Their relation is forward propulsion of progress. The expansionist critique of neo-Malthusianism highlights this absence of contradictions in the expansionist thesis.

From the expansionist perspective, neo-Malthusian finitude becomes necessarily a faulty premise which affects all levels of analysis including the level of methodology. As neo-Malthusians impute a particular finitude to nature's capabilities, this initial quantitative characterisation determines the boundaries of mathematical calculation. Incorrect calculations of nature's resources, partly derived from a narrow assumption of finitude, are mistaken as measures of real physical limits of nature. Accordingly for expansionists, the more accurate approximating estimates of the scale of nature, even in the restricted sense of earth's resources and capabilities, exceed those proposed by neo-Malthusians to a mind-boggling extent. Maddox depicts a planet of awesome dimensions.

> The atmosphere of the earth alone weighs more than 5000 million million tons, more than a million tons of air for each human being now alive. The water of the surface of the earth weighs more than 300 times as much – in other words, each living person's share of the water would just about fill a cube half a mile in each direction. The energy released in thunderstorms is comparable with that released by the explosion of nuclear weapons; hurricanes or typhoons are the equivalent of several thousand nuclear weapons. The momentum of circulation of the oceans and the atmosphere is vast.[5]

And again,

> The scale of the earth's oceans is a telling illustration of the durability of space-ship earth. With the present population of 3500 million, there is the equivalent of one tenth of a cubic mile of sea water for every person. There are more than seven million tons of gold in the waters of which the oceans are made, five pounds or thereabouts for every person now alive. The amounts of other materials in the oceans are much larger still. . . Sheer physical exhaustion of the resources of space-ship earth is obviously an exceedingly remote possibility.[6]

But, quite apart from the vastness of scale of nature's constitution and the enormity in the disparity between the real limits of the earth and those projected by the neo-Malthusians, the expansionists emphasise that limits cannot be conceived in absolute, physical terms. A mathematical characterisation of an object's dimensions is meaningless in itself. 'Smallness', 'limitedness' or 'shortage' are not absolute terms but economic notions and refer to particular socio-economic contexts. Food shortage, for the producers of food, peasants in the Third World for example, illustrates the socially determined rather than the physiologically determined sense of shortage. Maddox points to the social arrangements for food production and distribution as being determinant of whether or not food shortage can be vanquished. Food supply therefore is not directly a function of nature but rather of the mediating social relations. These are, for Maddox,

> arrangements for making sure that farmers in the developing world, now being blessed for the first time with food surpluses,

> will be able to win a just reward for the trouble. Nobody should think that these problems will be solved easily, but they are very different from the problems of absolute shortage with which most prophets of calamity are preoccupied. There is no doubt that the world could support 7,000 million people or even twice that number. The question is whether people will take the trouble to do so.[7]

Beckerman argues that the reason for the disparity between the real supply of resources and the claims and predictions of shortages is that the latter are 'economic' estimates being made by development companies who integrate the proportion of cost and private profit into the calculation. As he puts it,

> estimates of reserves at any moment of time never represent true reserves in the sense of being all that can ever be found, irrespective of the demand and the price. . . the known reserves represent the reserves that have been worth finding, given the price and the prospects of demand and the costs of exploration.[8]

There is another sense in the expansionist critique of the neo-Malthusian notion of limits. Shortage in relation to raw materials, for example, is false for it is static. It relates merely to a fixed quantity and ignores the capacity of scientific and technological means to alter it by discovering substitutes for a given resource. The continual technological innovations make the notion of shortage in the absolute sense meaningless by rendering it obsolete. Maddox predicts that

> in the world of the future, characterized by electronic computers and not the steam locomotives which gave the Industrial Revolution its special flavour, the use of raw materials will follow quite a different pattern from that with which the prophets are obsessed.[9]
>
> The idea that some materials play such an important part in modern life that civilisation would collapse without them is by all criteria an illustion.[10]

Beckerman rejects the argument of present scarcity and imminent exhaustion of resources for its failure to take account of a number of dynamic factors which operate in such a way as to reinforce economic growth and broaden rather than narrow the basis of expansion. The major factor, of course, is technological innovation which may render economically obsolete various resources which presently appear to be in decreasing supply.

> One of the favourable feed-backs in the economic system has been technical progress. Sometimes this takes the form of just technical progress in mining or exploration, so that new reserves have been found when needed, or lower-grade ores proven to be economically useful. But it also often takes the form of the development of completely new materials, or new uses for old materials.[11]

The World Bank's *Report on the Limits to Growth* reiterates that the 'concept of resources itself is a dynamic one: many things *become* resources over time. Each century has seen new resources emerge. The expansion of the last hundred years could not have been sustained but by the new resources of petroleum, aluminium and energy. What about the possibilities of tomorrow – solar energy, sea-bred resources, and what else'.[12]

As, for Beckerman, 'man is the measure of all things', it is his reason which imparts on nature, even as a life-sustaining solar system, the particular qualities it may possess at a given time. As long as he develops the technical means requisite to assert his mastery, no problems such as scarcity or exhaustion can arise. Thus the characteristics of nature are never fixed, absolute, static, for nature is fashioned by man's activity. To the extent that man is inherently rational, innovative, enterprising, industrial and imaginative, nature, rather than waning or burning up, expands and grows along with man. Accordingly, any natural resource is meaningful in and through man's activity. Beckerman illustrates this in the following example of 'radical changes in techniques of extracting or refining materials from ores that have enabled lower and lower grades of ore to be handled economically, such as the reduction in the lowest grade of copper which could be handled economically from about 3 per cent in 1880 to almost 0.4 per cent now.'[13]

And neither does the neo-Malthusian theory of demographic growth escape expansionist critique. Indeed, in the expansionist view, the entire theory, from its assumptions to its propositions of rates of demographic expansion to its predictions, is untenable. The Malthusian relationship between population and food production is faulty in that it holds that populations naturally grow at an exponential rate and food at an arithmetic rate making for an imbalance which inevitably leads to a number of possible catastrophic results. The characteristics attributed to both factors, namely population growth and food production, are false, a matter which invalidates also the relationship between them. Using patterns of demographic behaviour of industrialised societies as counter-evidence to the neo-Malthusian population thesis, expansionists contend that populations do not expand at an exponential rate particularly under conditions of sufficient or abundant material means. The experience of industrialised societies shows a pattern not of exponential increase but of decline and even that of the Third World indicates a rate of growth which, though higher, is far from being exponential. Thus the assumptions are simplistic and bear little relation to the complexity of social, economic, cultural as well as technological factors which are determinate in human reproduction. John Maddox accuses neo-Malthusians such as Paul Ehrlich of conducting an abstract, mathematical 'numbers game' with human sexuality.[14]

On the other hand, food production has, is and is likely to increase at higher rates than those projected by neo-Malthusians. Technological developments as well as the rate of agricultural productivity are increasing. Maddox gives a dramatic example of increasingly accelerated rate of growth of agricultural productivity. 'Since the agricultural revolution of neolithic times, the amount of land needed to support a single person has decreased from some tens of square miles to roughly an acre, a ten-thousandfold improvement, but there is still a long way to go before the productivity of agriculture bumps up against the theoretical limit, food enough for one person from each five square yards or so of the surface of the earth.'[15]

A positive assessment of technology underlies all these statements which lead to the contention that technological developments have made possible the increasing egalitarianism in the process of history. John Maddox points to various developments made in the last few decades in medical technology, transportation, telecommunications and various other fields which have been beneficial in that they increase the individual's liberty and improve his quality of life.[16] Technology is never detrimental in itself. Even in its most negative aspect, in its possible pollutive effect on the environment, it still retains its positive, constructive qualities for it has the capacity to provide the solution to problems. This pragmatic sense in which the production of a problem creates conditions for its solution is evident in Herman Kahn's defence of technology.

> The key point is that the ability to handle the problems of the future will be aided much more than hurt by increasing the world's wealth and technological capabilities. Such increases will permit more flexibility and effectiveness.
>
> It seems paradoxical to many people that the very problems that have been created by wealth and technology can also be solved by them, but this is not a genuine paradox. When one has wealth and technology without worrying about how to use them, problems are created. But the moment that serious concern arises over their actual and potential uses, the problems can usually be alleviated or prevented.[17]

But in any case for expansionism, problems arising from technology are not normal processes. That is, the normal process of development of technology need not, in itself, entail the generation of problems.

Neo-Malthusianism does not retain the same optimism regarding technology as a problem solver. Indeed, unrestrained technological expansion can itself generate socio-ecological ills. In its tendency to expand exponentially it hits against the absolute limits of the planet's capabilities and tends to threaten ecological viability. This position regarding technology, however, is not fundamentally antithetical to that of expansionism. The appeal is for a moderation of technological expansion within the finite boundaries of the earth's capabilities; it is not a plea for a replacement of present technology by some other form of technology.

These positions on technology are echoed in the place accorded to individualism and competitiveness in each respective system. In neo-Malthusianism, as we saw, individual competitiveness is depicted as excessive abandon. In expansionism, however, it is a measure of progress and a condition of freedom and liberty. Even social disparity acquires a positive value in expansionism where it is seen as constituting a stimulus for the deprived individual to seek to acquire, through his own devices, that which has already been achieved by the more successful. The riches of one individual provides for another, poorer individual the drive to achieve as much or more for himself. It creates the conditions for generating and sustaining individual incentive to compete, to acquire, to expand his possession. Competition is thus conceived as being the form of freedom with which the individual seeks to achieve his goals.

The same logic is applied in the expansionist defence of social disparity between rich and poor nations and in a proposal for a competitive form of interrelation between them. For Herman Kahn, for example, disparities between rich and poor countries propel expansion forward by sustaining competition.

> The increasing disparity between average incomes in the richest and poorest nations is usually seen as an unallowed evil to be overcome as rapidly as possible through enlightened policies by the advanced nations and international organizations. If this occurred because the poor were getting poorer, we would agree, but when it occurs at all, it is almost always because the rich are getting richer. This is not necessarily a bad thing for the poor. . . In contrast, we view this gap as a basic 'engine' of growth. It generates or supports most of the basic processes by which the poor are becoming rich, or at least less poor. The great abundance of resources of the developed world – capital, management, technology, and large markets in which to sell – makes possible the incredibly rapid progress of most of the developing countries.[18]

For Kahn, attempts by the state to reduce disparities by means of social welfare programmes can only have negative consequences on growth, for they reduce private incentive. 'The direct budgeted cost of welfare and social justice. . . could slow down economic growth and technological advancement. . . We focus on some of the more subtle psychological and cultural considerations. The first occurs when "everybody feels that the state owes him or her a living" (or knows it will ultimately provide), or when relief and welfare compete with private industry in economic incentives. . . it reduces motivation and the commitment to work and to succeed.'[19]

This same individualism as we see below, underlies the political theory of expansionism.

EXPANSIONIST POLITICAL THEORY

The classical liberal relationship between the private and public spheres remains essentially unchanged in expansionist theory. The public, political sphere is derived from the nature and exigencies of the private sphere. That is, private appropriation alone is the vehicle of human happiness, development and fulfilment. It is in and through the activity of private appropriation that man is free, creative and expressive of his full potential. The public sphere thus has no *raison d'être* other than protector of the private sphere. This assumes the form of a political apparatus which secures and preserves conditions for a viable and flourishing private sphere.

In classical liberalism this political apparatus is the state whose activity is dictated by the requirements of the private domain. As long as private appropriation proceeds relatively smoothly and unobstructed, the state simply provides the legal framework in which the interacting private individuals pursue their private interests in recognition of mutually binding contracts into which they, themselves, freely agree to enter. Such state-sanctioned contracts are simply expressions of unhampered private appropriating activity. Depending on the requirements of private appropriation, the state may embark on more active programmes such as providing the necessary material infrastructure of transportation, communication and, in various cases, sufficient irrigation works projects in order to secure appropriate conditions for private appropriation.

In general, that which distinguishes classical liberal political theory from neo-liberal political theory is that for the former, private appropriation is best served by a non-interventionist state whereas for the latter, private appropriation itself necessitates a more active interventionism. Expansionism is clearly in the tradition of classical liberalism in regard to the character of the state in its relation to the private individual. Kahn favours a more classical liberal state which restricts its economic intervention in economic growth to a minimum. 'We presume that far and away the safest, most normal and reasonable path is to go along with what might be thought as the "natural" path, where natural forces do the slowing down not government programs.'[20]

The Third World, however, is not put into this mould of passive public sphere, active and dominant private sphere. For expansionists such as Maddox, passive state is not entirely appropriate for the Third World. Here, active 'interventionist' state is more favourable to capital accumulation.

> Governments in developing countries. . . [play] an active part in the management of the transformation. Irrigation works must be built, responsibility has somehow to be accepted for the attestation of new varieties of seed and the training of the people who will use them. Fertilizers must be manufactured or imported. And it is clear, as the harrowing experience of East Bengal in 1970 indicates, that the existence of an agricultural market requires roads for carrying the crops away from the land.[21]

Both of these forms of state are compatible with individualism: the passive state is compatible with competitive individualism and the Third World interventionist state is necessary for creating a material infrastructure in which competitive individualism can emerge.

The individualism of this political theory is clearly highlighted in the expansionist conception of the proper resolution of environmental issues, such as pollution control. The individual applies the same cost-benefit calculations to selecting a means of pollution control as he applies to the buying and selling of commodities. He does so as sovereign elector in his choice of alternative political party programmes.

The question of pollution is not, for expansionists, a cause for universal alarm regarding irreversable contamination of the earth's resources, but a question of how much pollution the individual is willing to tolerate at what price. For, in the final analysis the individual also foots the bill for the extent, scope, and quality of the anti-pollution programmes by means of the taxation system.

Beckerman proposes the notion of 'optimum' as a measure of balance between costs and benefits of a given programme. In the case of pollution, the economically sensible choice of means for handling it, Beckerman contends, is 'optimum' pollution as 'the level at which the social costs of reducing pollution by a further unit just equal the social benefits of doing so, and where a further reduction in pollution would then cost more than the further benefits to be obtained from doing so.'[22] The assumption here is that the rational individual will choose 'optimum' alternatives, rationality being conceived purely in terms of the cost/benefit ratio. It is the individual who defines his values, determines his priorities and accordingly calculates the cost/benefit factors in pursuing them.

Expansionism also vests in the individual the right to select between various possible technologies and to determine its (or their) regulation. A need for control of technology may arise in such areas as medical and biological technology including genetic technology where various discoveries could lead to ways of possible control of human genetics. Since, in expansionism, the individual is sovereign in his choice of the modality of application of technology which he politically expresses in the liberal democratic electoral system, technology is always contained, for it is nothing other than a measure of the mastery of the individual over his world.

EXPANSIONIST ANTHROPOLOGICAL ASSUMPTION: THE PRIVATE INDIVIDUAL

The expansionist cast of technology, individualism and competitiveness as the keys to freedom and historical progress expresses the particular sense of the Hobbesian anthropological tenet underlying the whole expansionist construction. This tenet assumes not only that man is subject of determination but that this subject has the particular form of private individual. This means that the subject is not just any concrete individual but the bourgeois individual. The world he creates is in the image of his private subjectivity, namely a world of private property. The infinite appetites of acquisition which man is posited as having by nature are those of private property.

The activity of private appropriation mediates the development and expansion of the subject's world. It is in the light of the subject as private individual that we can understand the expansionist conception of nature as a network of organic causality which, empty of objective value remains open to the private individual's formative activity of private appropriation, nature expanding in proportion to his activity and thus defiant of any quantitative limits and measurable finitude. It is in this light that we may understand that which, on the surface, appears as an insensibility to ecological concerns relative to other forms of environmentalism, as for example, neo-Malthusianism.

This reading of expansionism as a form of Hobbesian liberalism points to the sense in which there is an affinity between expansionism and neo-Malthusianism in spite of the controversies between them. The affinity lies on the very fundamental level of anthropological assumptions about human nature. We find in this assumption the common conception that the subject of determination is the private individual who remains unconstrained in expansionism and becomes moderate in neo-Malthusianism. But the neo-Malthusian conception of moderation of the private individual is not a quality which inheres in the individual. On the contrary, it is externally imposed by a political system which sets quantitative boundaries on activity receiving its cues from ecological exigencies as the state of science is able to assess at any given time. For both expansionism and neo-Malthusianism the private individual has infinite, insatiable appetites and in his search to express and satisfy them he does not naturally seek a moderate course. The private individual is, if anything, naturally immoderate.

The difference between expansionism and neo-Malthusianism thus is not a difference of conception of human nature but rather a difference of judgement of the effects of a free reign of human nature. For expansionism the effects are wholly positive. It is through unconstrained competition between individuals who determine their goals and pursue them with a minimum of external impediment that progress is achieved. This is the Jeremy Bentham formula. For neo-Malthusianism on the other hand, the effects of the unconstrained activity of the private individual are potentially negative. But these negative effects are seen not so much as effects on social relations but as effects on nature. Unconstrained competitive activity generates pollution, exhausts natural resources, disturbs ecological balance, and generally threatens global planetary viability.

Part II
Eco-development

3 Theory of Eco-development

The term 'eco-development' was coined by the Secretary General of the 1972 Stockholm Conference of the Human Environment, Maurice Strong. He used it to mean an alternative form of economic development to the present pattern of economic expansion. In its present usage it refers to more than an economic theory. It is also a social and political theory associated with such names as Ignacy Sachs, Johann Galtung, Howard Daugherty, Charles Jeanneret-Grosjean. It is espoused by many economists and developers associated with various organisms of international organizations, as for example, the UNEP/UNCTAD Symposium, Cocoyoc, 1974, the UN Conference on Raw Materials and Development, New York, 1974, and research centres affiliated with international organisations such as the Centre international de recherche sur l'environnement, Paris, and the International Institute for Environment and Society, Berlin, as well as some state governmental departments such as Environment Canada, CIDA, and the Science and Technology Council of Mexico.

The literature of eco-development, however, is not of as theoretically elaborate a form as that of neo-Malthusianism nor that of radical ecology which we shall examine later. If eco-development is assessed on the basis of theoretical elaboration and articulation, it does not fare highly. It is not intended by its exponents as a general economic, political and social theory depicting a global economic, social and ecological malaise, explaining its sources and proposing a programme of economic, social and political reconstruction. It is more a collection of essays and research papers with a thematically narrow and methodologically empirical focus, whose theoretical interconnection is often far from evident. The themes include, for example, given economic programmes of one or other Third World social formation in a given time period and its economic and ecological effects. But this predominantly narrow and empirical focus of the literature as opposed to a more general and theoretical one, does not imply that the texts are theoretically disparate nor incoherent. There is, on the contrary, a basic logic that runs throughout this literature and distinguishes this thinking from other perspectives on ecological issues. It is this logic which we shall attempt to capture in this chapter in reconstructing the theoretical problematic of eco-development, the sense of its constitutive notions including ecology, the relation between the problematic and the programme of social, economic, political and ecological reconstruction.

Eco-development is a critique of economic concentration and political centralisation on the global level and above all a programme of economic and political decentralisation. The critique of economic and political centralisation is a critique of the effects of concentration rather than its causes. The sense of the critique thus is that centralisation of economic and political power distributes individuals, groups and social formations on a global level along a poor-rich dichotomy. The rich are those with economic and political power and the poor are those without it. That which guides the activity of the rich is accumulation, expansion, and that which guides the activity of the poor is subsistence. But social inequality is not the only effect of concentration. Another is environmental deterioration or ecological disorder in the form of 'possible exhaustion of nonrenewable resources, declining marine protein resources, the contamination of food chains by agricultural and industrial chemicals, and the potential disruption of the heat balance of earth/atmosphere energy system.'[1] Such an assessment of the economic and political centralisation leads to the proposal that it is socially and ecologically sensible to decentralise such that the rich-poor dichotomy is replaced by equality and that a condition of ecological viability is restored.

This is a very general statement of eco-development. It does perhaps little more than define a problem in tautological terms. Economic and political concentration is evidence of a have/have-not dichotomy and the have/have-not dichotomy is evidence of economic and political concentration. Since such a division is ecologically destructive as well as socially unjust a solution must involve replacing it by its opposite. Centralisation must be replaced by decentralisation, in the form of economically self-sufficient communities, or social formations which decide objectives and means of achieving them without external interference or control.

THE PROBLEMATIC OF ECO-DEVELOPMENT

While this simple formulation captures something of the spirit of eco-development it cries out for clarification. How can the question of economic concentration and political domination be examined without reference to the logic of the particular organisation of economic and political practice of which economic concentration and political domination are themselves the effects? How can this question of economic and political concentration be posed solely in terms of social and ecological effects? How does this manner of posing the critical questions affect the understanding of the object? How does this understanding affect the programme of social, political, economic and ecological reconstruction? It is the problematic of eco-development that may provide clues to the answers of such questions and indeed may reveal something of the internal logic of the whole theory of eco-development.

The critical notions of the problematic of eco-development are production, consumption, ecology, basic human needs. Production is posed as an intentional activity for need satisfaction and consumption is posed both in terms of the use of a produced object as transformed nature and in terms of the use of nature for producing usable objects. The first and most common notion of consumption conveys or, at least, implies a distinction between production of a product and the use of the completed product as its final and terminal destination and retains only the latter moment. The second notion of consumption does not distinguish either implicitly or explicitly between these two moments. It rather collapses them into one and consumption becomes interchangeable with production in the normally understood sense of a work process comprising human labour, tools and raw materials for producing usable, consumable products. It is this second sense which dominates or prevails over the first in this problematic.

How does this notion of consumption as a combination of consumption and production reflect the logic of the problematic and affect the sense which this problematic accords its other constitutive notions and the questions it poses? First, the collapsing of production and consumption into a single, undifferentiated activity reflects the specific logic of the problematic which situates this activity less on the level of social relations and more on the level of a direct relation with nature. Nature is conceived as being a directly consumable object and a raw material transformable into usable objects. In both cases consumption relates to the activity of 'using' or consuming nature without there being a distinction between the moments constitutive of this activity. Ecology is the state of nature as the effect of this activity. In this problematic, social relations are seen to be the effect of the relation with nature such that individuals, groups, and social formations are cast in a rich-poor dichotomy as less the effect of determinate social relations and more the effect of a direct relation with nature, or in other words, the effect of consuming nature. The rich are rich by virtue of a particular way of consuming nature, the poor are poor by virtue of another way of consuming nature. Similarly, ecological imbalances are not the effect of socially determined activity but rather the effect of technical manipulations of nature, i.e. technology. Both the form of consumption of nature specific to the rich and that specific to the poor generate ecological imbalances though not of the same type. They are both excessive but this excess assumes a different form generating correspondingly different ecological disorders.

This problematic of consumption that poses the question of rich and poor social differentiation as a critique of patterns of consuming nature. The consumption pattern of the rich is characterised by an excessive waste and spoilage of the environment and that of the poor by excessive use or abuse of natural resources as, for example, over-utilisation of land for specific agricultural purposes leading to erosion or soil depletion. 'It is widely realized today,' Daugherty *et al.* argue, 'that these increased social inequalities have led toward a dual degradation of the environment by: (1) the small wealthy elite, wasting scare resources in ostentatious consumption patterns (for which a very large and still growing part of the total national income has to be allocated); and (2) the poor, largely majoritarian segment of the population, overusing by necessity the few natural resources to which they have access.'[2] Thus, the rich and poor owe their conditions to an equally faulty pattern of consumption as both exceed quantities compatible with ecological exigencies. In the case of the rich, excess refers to the quantity of appropriated nature over and above that which is necessary. In the case of the poor, excess refers to the quantity of appropriated nature necessary for survival but which, nevertheless, induces ecological imbalance. Referring to the specific ecological distribution of the earth's natural resources, Ignacy Sachs states the effects of the patterns of use of these resources in this way.

> The paradox of the world in which we live is that bad distribution of wealth provokes wastage on both poles of society at the same time. The wealthy overconsume and drain away the great majority of available resources; they do so by misusing vast spaces of potentially agricultural land to boot. The poor underconsume and, driven by poverty, overuse the rare resources to which they have access. The struggle against waste thus appears to be inseparably linked to the struggles against poverty and against bad management of the environment.[3]

Consumption is conceived in terms of direct appropriation of nature, a notion which serves to define social inequality as a disparity between the rich and poor drawing from their differential degrees and form of appropriating nature directly. In this sense, social inequality may be assessed as a measure of disturbance of the ecosystem although the converse relation between ecological damage and social inequality may not necessarily hold. That is, the degree of disturbance of the ecosystem may not be a measure of the degree of disparity of social inequality. For, if a simple redistributive system is introduced to reduce social inequality as differing excesses in consumption, the ecological balance is not necessarily restored. As long as the total appropriation or consumption of nature remains at an excessive level simple redistribution to reduce social inequality does not resolve the ecological imbalance. The theory of eco-development, thus, rejects the Soviet model and similar socialist models of social organisation and production on the grounds that a centralised, state accumulation is as destructive of the environment as private accumulation of western systems.

> The countries of Eastern Europe, apparently better equipped institutionally to apply a voluntaristic policy of ecological caution, have not taken advantage of the situation, preferring rather to sacrifice all as quickly as possible on the altar of economic expansion. In regards to the use of this expansion, their performance is hardly more imaginative in the sense that they are simply content to recreate a consumer society similar to that of the western model, the only difference being that the narrower gap in income opens up the possibility of a petit bourgeois life style to a more sizeable fraction of the population.[4]

The problematic of consumption highlights technology as the source of social and ecological ills. This problematic implies a process of production as the technical process of working the raw materials of nature into a usable or consumable product. Consumption here is really the moment of technical mediation between producer and nature. It is conceived as being a relation between producer and nature but not one, it would appear, which is, itself, mediated by a definite and determined social relation. Consumption or the technical utilisation of nature appears, rather, to be a direct, socially unmediated relation. The properties of the direct producer derive, in this thinking, from the relation he sustains with nature and the quantity and quality of products he extracts from nature. His being poor owes to a particular type of consumption, in this sense of working nature in a particular way, such as by primitive means of tilling the soil, and to a particular degree – over and above the soil's tolerance. His being rich owes to another type of consumption such as applying the productive technics of heavy industry and high technology to 'working' nature. Poverty and richness thus appear to be the simple effects of forms of technology as a direct relation with nature.

But the notion of technology which is posed as the source of ecological disorders and hence also the means of restoration of ecological balance and harmony refers to more than tools and their manipulation. It includes also a technical division of labour as a form of social organisation for the application of tools on raw materials. It does not include division of the product and hence excludes social relations in this sense. This notion of technology is consistent with positing production as an undifferentiated moment of consumption in which producers conduct the work process as the beneficiaries or consumers of the product. The division of the product hence is not 'problematised' by the problematic of consumption. As a work process by direct producers, individually or as a group, technology contains a subjectivist element for what sets technology in motion is a positing of objectives. But this positing of objectives can be either a rational or an irrational act depending upon its final effect on ecology and society. It is an irrational act if the objective is excessive production with the effect of setting in motion productive technics of heavy industry and high technology despoiling nature beyond its tolerance capacity. It is also irrational if the objective is simply subsistence production where primitive technics are applied in ways abusive to ecological requirements. These are two different forms of irrationality, the former specific to the rich, the latter to the poor. Both, however, have negative effects on ecology by threatening ecological viability. They also have negative effects on society by reproducing a rich-poor dichotomy. The rationality of the act of positing is similarly a function of the objective and its effect on ecology and society. It is rational if the objective is both moderate production as opposed to excessive and if it sets in motion a technology compatible with ecological balance and social equality and hence with positive and constructive ecological and social effects.

Now, what is the basis of this rationality? Is it the knowledge of natural laws governing ecological processes? Is it a scientific knowledge, hence, obtained from the method of science? Is it a knowledge from which can be derived a scientific stipulation of ways and means of relating to nature that do not exceed natural organic tolerance? Depending on the level of development of science such a knowledge may suffice to avert ecological disasters or even minor ecological disorders but how could it resolve the rich/poor dichotomy by replacing it with social equality?

ANTHROPOLOGY AND BASIC HUMAN NEEDS

Although knowledge of natural processes is a component of this rationality there is evidence in the theory of eco-development that it is neither the sole nor the major constituent of rationality. Rather, this rationality is an anthropological rationality of human essence. It is the theory of basic human need which provides this evidence. Basic human need is conceived both as human essence and as a regulatory device of consumption of nature. The act of positing objectives, the objectives and the means of achieving them are tied together in the notion of basic human need. It is basic human need which makes them rational. The act of positing objectives is the moment of free subjectivity as an expression of human essence; the objectives posited, i.e. production for basic human needs, are the form of this human essence; the means of achieving the objectives and hence the satisfaction of basic human needs is realisation of human essence.

The sense of 'basic' in this notion of basic human need refers to more than vital biological needs. It includes a sense of 'subject' as creativity. It is basic to human nature to be free, to create, and to determine the form of this creativity. Hence the essential quality of the needs is that they be determined by the 'needer'. It is in the act of positing needs that needs are created, making the human being the *subject* of needs not subject *to* needs. Human reason, thus, defines needs. Indeed, the various definitions of basic human need highlight this element of free subjectivity while retaining an allusion to biological needs, the sense being that free subjectivity determines the form of satisfaction of biological needs. The authors of the Cocoyoc Declaration provide a list of basic human needs which includes 'health and education, freedom of expression and impression, the right to give and receive ideas and stimulus, the right to work at a self-fulfilling work, the right not to be alienated through production processes that use human beings as tools.'[5] In his definition of basic human needs, Johann Galtung proposes not specific needs but a methodology for arriving at basic human needs in which the subject determines his needs. The method, which by his own admission is difficult or impossible to apply, involves canvassing everybody for all their wants and wishes at any time for the preparation of individual lists of needs, the overlap of which would provide a guide to universal basic human needs. The overlap is the result of individually distinct expressions of general human needs.

The exclusion of the possibility of obtaining such complete information leads Galtung to propose not a list of specific needs but a typology of a high level of abstraction of material and non-material needs – security (from individual and collective violence); welfare, including nutrition, water, air, sleep; movement and excretion, protection against climate, disease and strain; dialogue; and education, being the five principal types of material needs and a need for self-expression, work creativity, a need for self-actualisation, a need for happiness and freedom (expression, freedom to choose friends) being non-material needs.[6]

Basic human needs are created in the act of positing since they are the quality of the subject's freedom. In this sense, they are comparable to the Hobbesian expansionist insatiable desires and appetites as they are conceived as being essential to the subject and not as having a separate existence. That is, in neither case is the anthropology one which conceives man as an instrument or means of achieving externally posited goals. But the act of positing distinguishes the subject in the theory of eco-development from the subject in the theory of expansionism. The subject of eco-development is a moderate subject not an insatiable, hoarding subject. Because he is tempered, the needs he posits are moderate, even abstemious.

What emerges from the theoretical construction of basic human needs, is a one-stroke resolution of both social disparity between rich and poor and ecological damage. This is so since the free expression of basic human needs represents the proper consumption of nature and it is this which makes for ecological harmony deriving from the appropriate amount and appropriate mode of consumption. The overconsumption of the rich as rich and of the poor as poor which leads to ecological disorder is brought to a level of 'human' consumption. Basic human needs, thus, are at once consistent with the inherent needs or 'essence' of the producer and with ecological exigencies.

This anthropology of human essence is not that of private or 'possessive' individualism. The subject of this anthropology is not the private individual as in the case of expansionism, nor the modified private individual in the case of neo-Malthusianism. The subject, here, appears to be a collective subject, the eco-community or local community. It posits its basic human needs and develops a technology appropriate to the ecological peculiarities of the region it inhabits in order to satisfy these needs in conformity with ecological exigencies. But is this anthropology one of collective human essence comparable, for example, to the qualities of the Marxian 'species being'? If this is the case, an apparent contradiction would emerge between an anthropology of collective essence and the problematic of eco-development which poses a rich/poor dichotomy and ecological disorders as effects of a direct relation with nature hence as an asocial relation or a socially unmediated relation. Before examining this crucial question, let us reconstruct the economic theory of self-reliance and the political theory of local autonomy and global decentralisation, for they impinge directly on the question of local or eco-community.

AUTONOMY AND SELF-RELIANCE

Economic self-reliance and political autonomy are the characteristics which eco-development attributes to the local community or eco-community. What is the local community and what is the sense of these attributes? Is it simply the municipality as an entity of local jurisdiction over such matters as waste disposal, local transportation, municipal parks and so on? Is it the municipality as it is presently constituted within the political framework of a nation-state as the lowest of a multi-tiered system of political jurisdiction and, if so, would a simple inversion of the present system of state jurisdiction, which would concentrate power of self-determination at the local level, produce autonomous and self-sufficient local communities?

The local community of eco-development is not an empirically rationalised term as a referent to a municipality. It is not a formal political structure. It is not simply a politically constituted citizenry with the power of self-determination. It is not externally constituted as a community by a simple act of jurisdiction. Rather, community has an organic form. It has an internal coherence though not necessarily of economic, ethnic or linguistic homogeneity. It may, indeed, be socially pluralistic, composed of a multiplicity of groups with each distinct group pursuing its respective interests. But this social plurality is united by a common underlying orientation, a common sense of history and a common sense of destiny. It perceives itself as a whole community. It seems, thus, that the local community is socio-historically determined. It has as well, however, a geo-ecological determination in that it occupies a geographical space which constitutes an ecologically coherent system. These two determinations of local community serve to distinguish and differentiate one community from another.

According to the economic and political theory of eco-development, a local community's distinct social identity, in the subjective sense of its perceiving itself as a unique entity with a common history and a will, justifies political autonomy on the grounds of the democratic principle of self-determination. Its occupying a distinct geo-ecological space of some ecological coherence in which, with the appropriate technics of production, it becomes self-sufficient, i.e. an 'eco-region', is justifiable on rational, economic grounds. These two characteristics are not seen to be independent one from the other. Indeed, it is claimed that to the degree that political autonomy and economic self-sufficiency are interdependent, with self-sufficiency being a condition for real political autonomy, so is eco-production in a specific eco-region part of the common history and sense of identity of the local community. There are, thus, two distinct, mutually reinforcing dimensions here: the subjective, as a shared sense of identity of local community, and the objective, as the various objective socio-economic factors of the community's history. The latter creates conditions for the former, i.e. shared identity, and the former serves to reproduce these conditions.

The theory distinguishes between the qualities or properties of communities and the exercise of these properties. In the existing global economic and political system, organically constituted communities are not necessarily autonomous nor self-sufficient. According to the theory, communities, in the subjective sense of common history and common sense of destiny and in the objective sense of occupying a common 'eco-region', exist presently but they are not necessarily politically autonomous nor economically self-sufficient. This is seen to be the contradiction of our time for such local communities exist deprived of the means of self-sufficiency and real sovereignty. Eco-development is proposed, thus, as a programme for achieving autonomy and self-sufficiency.

> Eco-development is a style of development which, in each ecoregion, insists on specific solutions to particular problems in the light of ecological as well as cultural data, of immediate as well as long term necessities. It operates therefore with criteria of progress relative to each case where adaptation to the milieu as postulated by anthropologists play an important role. Without negating the importance of exchange, to which we shall return, it attempts to respond to the predominant mode of supposedly universalistic solutions and passe-partout formulas.[7]

In the perspective of eco-development these two criteria of community, i.e. common identity and common habitat apply to many or most Third World social formations which at present are neither self-sufficient nor autonomous.

The critical political question is how do such communities achieve self-reliance and political independence? The answer eco-development provides is situated on an economic not a political level. That is, independence is achieved by economic self-reliance hence the form of political liberation or independence is economic. This economic self-reliance is mediated by the appropriate technology. Consistent with the problematic of consumption of nature which poses the problem of social inequality and ecological disorder on the level of direct consumption of nature, this solution focuses exclusively on technology. It is one which combines the cultural peculiarities of the community with the ecological particularities of its habitat in the production of basic human needs. Ecologically, it is a means of harnessing from nature just that amount of energy necessary for production for basic human needs according to principles of conservation and adaptation to normal processes of physical and biological reproduction. Culturally, it is a means of production as historically developed by a community in its ecological habitat. In both these senses it is an eco-technic which may vary both spatially and temporally. What is appropriate for one eco-region and cultural community is not appropriate for another. What is appropriate today may not be appropriate tomorrow. This means that a primitive process of production is not necessarily more ecologically sound than heavy technology nor vice versa. Similarly, the eco-technic of production for this generation may not be appropriate for two generations from now.

Sachs defines ecotechnology as consisting of a synthesis of various technologies including a local adaptation of imported, foreign technology compatible with the local requirements of a local community occupying a specific eco-region.

> The valorisation of construction material of local origin, abundant and cheap – from bamboo to adobe – a relatively well-studied problem but where, in practice, much still remains to be done beginning with the rejection of an alienating value system which makes of an aluminum roof or of a steel and cement house imported at an enormous cost, the symbol of modernity, even in the bush.[8]

A particular eco-technology thus is a specific combination of various technologies, 'par rapport à un contexte historique, socio-économique et écologique donne' (in relation to a given historical, socio-economic and ecological context).[9] This implies that the local community expresses its creativity by selecting, combining and adapting particular technologies to specific local requirements.

> Besides, the concept of *self-reliance*, we have said, should not be confused with that of autocracy. There is nothing to prevent solutions involving a selected importation of knowledge, equipment and material. No country, however large, can pursue a technological policy without resorting to purchasing a certain number of 'black boxes', without adapting and concentrating its creative effort on well selected targets at the risk of an unfortunate dispersal of efforts and talents. What is important is that it be this country, and not its partners to the North, which decides its priorities and chooses its suppliers favouring cooperation between countries of the Third World at every possible occasion.[10]

SELF-RELIANCE AS TECHNOLOGICAL SELF-RELIANCE: POLITICAL IMPLICATIONS FOR 'COMMUNITY'

This conception of self-reliance contains the key to the notion of community in this theory. Self-reliance is the material reproduction of community as community. It is a form of communal liberation. But of what does this material reproduction consist? In other words, how does community reproduce itself as community? The question relates to the mode of reproduction of community. Specifically, it refers to the social relations constitutive of community and to the mode of reproduction of these social relations. But what can we say about the social relations of this community on the basis of eco-development's particular understanding of self-reliance?

Self-reliance is positing basic human needs and satisfying them in ways compatible with ecological order without recourse to external assistance or hindrance. But the focus of self-reliance is not on the actual act of positing basic human needs. Rather, the focus is on the means of satisfying these basic human needs for it is the means which confers freedom on the subject. This means of satisfying basic human needs is eco-technology as both technically exploitative knowledge of the operations of nature in their ecological manifestations, i.e. eco-science, and the application of this knowledge for the satisfaction of basic human needs consistent with the 'needs' of nature, i.e. eco-technic. Eco-technology is the technical exploitation of such knowledge for such a purpose to such an end. That is, it is the ecologically positive means of production for basic human needs. It is this technology which makes a community self-reliant and by the same token confers upon it its independence from others and hence its freedom. To answer our question regarding the relation of community and the mode of their reproduction, then, we must look to eco-technology.

Eco-technology, as we saw, comprises both a cultural and purely technical component and the whole question of the mode of reproduction of social relations of community revolves around the sense of these two elements of eco-technology. In a technical sense, eco-technology refers to the necessary technical operations for transforming a raw material into a 'needed' product. That is, it is technical labour producing consumable products for the satisfaction of basic human needs. In a cultural sense, eco-technology is the manner in which these technical operations are actually carried out. The actual labour process, then, is a function of both these factors.

Whether a given product is produced by one producer or a group of producers is determined by the technical complexity of the operations and the cultural norms of the community. The division of labour of eco-technology, thus, owes to both technical and cultural factors.

If this is an accurate depiction of eco-technology as constitutive of technic and culture, the cultural dimension simply specifies the manner of performing technical operations. It is restricted to the technical process of labour and as such differs little or not at all from the technical component of eco-technology. Culture as delineated by eco-technology says nothing about the subject of labour, neither the moment of positing the objectives of labour, i.e. basic human needs, nor the moment of disposing the objects of labour, i.e. the distribution of the social product. This eco-technological sense of culture abstracts the subjective moment of positing and disposing the product from the labour process. The cultural aspect of eco-technology is thus limited or reduced to the technical aspect and eco-technology is essentially technology.

We may note in passing that Schumacher's influential 'small is beautiful' thesis[11] has much in common with eco-development on the issue of technology. Relative smallness or decentralisation, accessibility, conservationism, adaptability are qualities of the Schumacherian 'intermediate technology' as they are of eco-technology. More than this, however, the new technology is central to social change in both conceptions. While neither the notion of eco-technology nor the notion of intermediate technology are limited to tools or machinery of work but extend to their application, they highlight the 'relation with nature' rather than 'social relations' implying that the former is determinant of the latter.

In terms of this sense of eco-technology as pure and simple technology, i.e. as productive forces, how can we answer our question regarding productive relations of community, also the mode by which the community reproduces itself as community? The obvious answer is that we cannot know the productive social relations since eco-technology signifies simply productive forces. We cannot know productive relations because we cannot derive them from productive forces. From this we can draw one of two conclusions: First, the theory of eco-development says nothing explicit about productive relations and hence it advances an incomplete theory of community. Since the full sense of community is not worked out, this task still remains to be done. The other conclusion, and the most plausible, is that the logic of eco-development collapses productive relations into productive forces. It theorises a labour process in terms of a relation with nature as one not socially mediated by social relations. The implication for community is that it is an entity which reproduces itself technically. This conclusion is consistent with the problematic of eco-development which poses questions of dependence, inequality, ecological disorder in terms of 'consumption of nature'. In this second conclusion, the logic of eco-development could not allow a more comprehensive sense of community which includes productive social relations.

If the second conclusion is retained we are confronted with a paradox. The community of eco-development is a non-community or acommunity since its mode of being is technical and the mode by which it is reproduced is eco-technological. The model of such a community or rather acommunity is the solitary individual in direct exchange with nature as a means of self-sufficiency which, as we shall see, is the model underlying radical ecology. If this is so, the anthropology underlying eco-development is an anthropology of individual self-sufficiency rather than a collective anthropology. The community is not a collective subject as one which posits social objectives, realises them collectively and determines the disposition of the social product collectively. Rather it is a plural entity as solitary individual or an aggregate of solitary individuals. If this is a correct interpretation it settles the issue that we raised in the last section regarding a possible contradiction between a collective anthropology and a problematic of consumption of nature. There is no contradiction for there is no collective anthropology here.

If this interpretation is correct, how can we explain a theory of community which turns out to be a theory of acommunity? Can we simply say that it is a gross error which invalidates the theory and leave it at that? Or can we find an explanation of this paradox which does not reduce it to a simple error as insufficiency. Theoretical paradoxes and contradictions are never 'innocent' errors. They are determined by the internal logic of the theory – a logic which has its own basis of determination. It is a political basis and hence the sense of these 'errors' is disclosed on a political level.

One clue of this political basis is provided by the problematic of eco-development which poses capital concentration as a problem in regard to its effects (i.e. social inequality and ecological disorder) and not its causes. This, itself, is not an innocent oversight for the cause of concentration, i.e. law of expansion immanent in capital or self-expanding capital, cannot be implicitly retrieved by the theory of eco-development. That is, since relations of private property are left outside the scope of the problematic of eco-development, they are not posited as constituting a problem and hence by implication they are assumed to be external to the crisis of ecology. If they lie implicitly outside the scope of the problem they are implicitly not external to the scope of the solution, namely a community of self-reliance. The solution of self-reliance is not a 'neutral' solution, or at least one equally favourable to all members of the 'community'. Rather it favours some to the detriment of others and in this sense it is political.

The community theory of eco-development leaves untouched the capitalist relations of production in posing community as achievable without a social, economic, and political revolution. Such a 'community' is one which itself, is marked by contradictory relations between capital and labour. This 'community' has an explicit empirical referent. It is the Third World social formations whose road to liberation is independence from the centre and assumed to be achievable by self-reliance. But this self-reliance is posited in terms of technology – a conception which leaves unaltered the prevailing and dominant capitalist relations of production of these social formations. As such, it is a class solution not a universal solution of a socially undifferentiated community in respect to class. This class solution is not one of the capitalist class of the centre for indeed it is from this class that independence is sought. Rather it is a solution of the dominant class of the periphery promoted as a solution of the whole undifferentiated community. But as this leaves intact the internal social relations it is a solution which favours the dominance of the local dominant class. The political aspect of eco-development lies in this relation between a programme of technological self-reliance and class dominance of the dominant class.

Part III
Radical Ecology

Radical socio-ecological theory represents a dramatic departure from the premises and assumptions of the socio-ecological theories we have examined so far. This departure is not evident in the conception of ecological deterioration as socially induced ecological destruction nor in the salience of this issue. Ecological deterioration is highlighted in neo-Malthusianism and in the theory of eco-development. The differences are of a more fundamental nature. They are differences of premises and assumptions – differences reflected in the very conceptualisation of the theoretical problematic. In neo-Malthusian theory and in the theory of eco-development, this problematic is a problematic of consumption of nature. Consumption is posed as a relation between a whole consuming unit in the form of a socially undifferentiated whole population in neo-Malthusianism and a community in the theory of eco-development. In the problematic of neo-Malthusianism, consumption is qualified quantitatively as a measure of natural organic tolerance, and in the theory of eco-development consumption is a measure of basic human needs and natural tolerance.

Radical socio-ecological theory alters the problematic of social reproduction general to social theory in a more radical way than does neo-Malthusianism or the theory of eco-development. Its problematic poses a relation between man and nature not as one of social consumption of nature but rather in terms of a relation of co-operation with nature where, in some theoretical constructions such as that of eco-anarchism, nature itself acquires subjective constitution. The radical quality of this problematic lies in its posing a relation between man, a human subject, and nature not as one mediated by social relations. Rather, the relation between man and nature is posed as being itself a social relation between two collaborative, communicating subjects, since nature assumes the properties of subject. Indeed, the problematic excludes social relations as anything other than the relation between man and nature.

We propose the hypothesis that an anthropological premise

underlies and determines this problematic and radical socio-ecological theory in general. This anthropological premise differs from that of neo-Malthusianism and from the theory of eco-development which we argued is a liberal premise assuming private, bourgeois individualism to be immanent in man's nature. It differs radically in two senses; the anthropological premise of radical ecology is neither a premise of private individualism nor a premise of social individualism. In the theoretical construction of radical ecology the human being is attributed a reason which unfolds in a direct relation with nature. A 'self-activity', a notion clearly highlighted in eco-anarchism, mediates between reason and nature. It is the form of development of man's subjectivity. We suggest that the premise of radical ecology is an anthropology of asocial self-sufficiency where self-activity is one of organic, socially unmediated exchange with nature. It is this hypothesis which reveals the theoretical affinity between theories which appear, on the surface, rather disparate and which indeed claim different and even conflicting theoretical inspiration. These theories represent the thinking of the broad spectrum of the radical ecological anti-nuclear peace movement including, and in particular, the alternative movement.

In order to explore this claim that radical ecology is fundamentally united however disparate the movement may appear to be on the surface, and that this unity derives from the shared common assumption of human nature as asocially self-sufficient, we shall examine three distinct theories: those of Murray Bookchin, André Gorz and William Leiss. Our choice is motivated by two factors. First, they are appropriate as illustrative cases of radical ecology both in terms of their influence on the movement and in terms of capturing the spirit of the movement. Secondly, insofar as each claims a distinct and differing theoretical inspiration – Murray Bookchin anarchist, André Gorz Marxist, and William Leiss liberal – they serve particularly well to test the tenability of our hypothesis.

4 Theory of Eco-anarchism: Bookchin's Critique of Authority

Bookchin's sociological analysis is a social critique. Its focus is authority. In our reading of his critique of authority, it is not simply a critique of relations of subordination and superordination in which one has the right to command and the other the duty to obey. It is, rather, a critique of social relations as such. This is the interpretive hypothesis we suggest for capturing the sense of his social critique. Bookchin perceives all relations of authority as relations of domination which could be formalised, in his understanding of them, as hierarchical relations of inequality in which A dictates the action of B and B submits to A's directives. Such relations are necessarily social relations.

That a relation of authority is social is, of course, true even by definition. There cannot be command and obedience where, on the one hand, there is no commander and, on the other hand, no obeyer. What is significant in Bookchin's critique of authority is not the proposition that all relations of authority are social, but rather that all social relations are relations of authority. In his critique, authority is necessarily social and the social is necessarily authoritarian but neither authority nor the social are ontologically necessary. That is, neither authority nor the social is a necessary condition for man's being and, more importantly, neither is a necessary condition for the expression of human essence.

If such a proposition does indeed underly Bookchin's theory of authority it could only be sustained if it is assumed that man possesses an essence and that such an essence is an asocial essence. Bookchin's work provides ample evidence of such an assumption – evidence which often comes in the form of the unstated or the missing. The most revealing 'unstated' form of evidence is the absence, even by implication, of an ontological theory of social determination. For if social relations are relations of authority and therefore transcendable into something else, the whole point of his critique, they are not ontologically necessary but rather historically contingent. His quarrel with Marxism, for example, is based on his rejection of forms of economic determinism as an account of social domination. Contrary to Marxism, Bookchin contends that a relationship of domination is not one which necessarily derives either in the first or last instance from productive relations of classes to the means of production, for it is not restricted to a relation of property but extends to a relation between generations, sexes, within the family, church and so on, any one of which may be the dominant source of authority in society. He gives historical examples which he interprets as ones where

> Many other standards, often totally at odds with our own, were adopted – most notably, disaccumulation rather than accumulation, of which the potlatch ceremonies of the Northwest Coast Indians are an extreme example. Even if we look beyond tribal life to more politically organized societies, we witness an orgy of mortuary construction and the rearing of lavish public buildings of which Egypt's pyramids and Mesopotamia's ziggurats are extreme examples of another kind. Conventional theories based on class analyses to the contrary notwithstanding, rulership rests less on proprietorship, personal possessions, wealth, and acquisition – in short, the *objects* that confer power – than it did on the *symbolic* weight of status, communal representation, religious authority and the disaccumulation of goods that the Neolithic village had hallowed.[1] (author's emphasis).

It would appear then that in Bookchin's conceptualisation, social domination is not necessarily economically, politically, or generationally determined. Rather, all factors could prevail in a given society making the actual form of domination an entirely historically contingent matter. He does not situate ideological determination however, in the same category as historically contingent determination. It falls rather in a category of socially necessary determination, such that while all relations of authority or domination are necessarily social they are also necessarily ideological or ideologically mediated. This ideological mediation of social relations, in Bookchin's critical assessment, has the function of specifying, as well as justifying, the hierarchical structure of status and the corresponding privileges and duties.

The notion of ideology underlying the proposition of a necessarily ideological mediation of social relations provides a clue to Bookchin's anthropological theory of human nature as asocially self-sufficient. To facilitate our seeing this we must continue our examination of Bookchin's understanding of social relations as necessarily authoritarian and necessarily ideological. For Bookchin, a social relation is objectively a relation of authority but a part of this objectivity derives from the ideologically mediated subjectivity of the actors in the relation. They share a common perception of privileges and duties. This, and not merely physical force, is what sustains a relation of authority. This is the reason, as we have already noted, that the relation of authority is necessarily ideologically mediated but not necessarily economically determined. Bookchin merely claims a relation of possible correspondence between ideologically mediated authority and property relations. Ideology thus is determined not by a particular form of social relation, i.e., economic or political, but rather by social relations as such.

Although this notion of ideology rejects any particular social form of determination but claims rather social determination in general and in this is opposed to the Marxist tenet of economic determination in the last instance, it does retain one sense of the Marxist conception of ideology. It is that ideology rewrites an objective social reality and makes it appear other than what it is. This proposition has led to the Marxist interpretation of liberal ideology as one which depicts a social reality of inequality as one of social equality. It would be an error to interpret ideology in Bookchin's schema, as converting a social inequality into an ideological equality for the objective social structure of hierarchy requires that those in subordinate positions comply with the directives of those in command and perceive themselves as having the duty to obey either because of their youth and inexperience, and/or their relative lack of expertise, and/or their lack of wealth, etc. The nature of the distortion which ideology performs is a concealment of what the actual social reality should be. This is not a concealment of an inversion type where, for example, real inequality appears ideologically as equality. This is a concealment of that which the social reality, in its very existence, denies. What ideology conceals is that social reality negates the true nature or essence of man. It prevents it from being developed and freely expressed.

What is this human nature to which Bookchin implicitly alludes in the various notions figuring in his critique of authority? The first characteristic of the nature of Bookchin's man which we should retain is that he is an auto-determining being rather than a socially determining individual being. Secondly, the process of auto-determination is a process which Bookchin would qualify as a process of self-activity and which we would interpret as a process of asocial self-sufficiency, for man achieves his essence, his true nature as free and autonomous, by relying on his own devices and by becoming self-sufficient in a radically asocial sense.

Bookchin is not, of course, unique in his construction of a theory based on the anthropological assumption of human nature, nor, for that matter, is Gorz nor is Leiss as we shall see later. An anthropological theory of human nature underlies a whole range of theories. In liberalism, for example, human nature is conceived on the model of the private, bourgeois individual. Liberal man, however, is essentially social, for his property which he appropriates is the labour of another. That he is a bourgeois individual owes to the relation of private property he sustains with labour, the actual producer of value. According to our interpretation, Bookchin's notion is of a free, autonomous and asocially self-sufficient individual, sustaining a direct relationship with nature free of any social mediation. The goals posited are ones of self-sufficiency and autonomy and the means developed to achieve them are not social relations, but technical skills necessary and sufficient to assure the individual's own reproduction as autonomous and self-sufficient without any dependence on alter. The independence of this individual is an asocial independence in an extreme and radical sense of negation of the social. Bookchin's individual defines his or her needs and goals and achieves them asocially.

We may say that the concealment which operates in Bookchin's notion of ideology is a concealment of asocial self-sufficiency. It would seem that ideology conceals human essence by situating the individual in a social relation of interdependence, necessarily antithetical to human nature, and making this social relation appear to be 'in the nature of things', i.e. to derive from human nature. Where should the human being properly be situated? That is, what kind of social formation would the free expression of human essence produce? Surely the unimpeded expression of human essence would not imply that the individual, asocial as he is conceived to be, reverts literally to a state of nature without any physical contact with other individuals. The absurdity of this is obvious. Neither, however, can a social formation be a society of structured social relations, for indeed, such structured relations would be supplanted. Rather, what Bookchin's notion of human nature implies is a society of independent, autonomous, asocially self-sufficient individuals. This society would simply be a sum of such asocial individuals.

If we are correct in our interpretation of Bookchin's critique of authority and domination as being really a critique of the social or of intersubjectivity from which we derive the notion of society as a sum of asocially self-sufficient individuals, how can Bookchin account for the emergence of the social in the first place? If, in other words, man is by nature asocial, how do social relations arise in the history of man's experience? Bookchin does indeed offer an account of this – an account which provides further evidence of his conception of society as a sum of asocial individuals. It is his theory of scarcity which contains at least a part of the account. It is a theory which is embedded in his socio-ecological analysis and critique of authority and which is possible to reconstruct from his various arguments regarding, for example, individual liberation, human needs and technology. In some respects it bears a similarity, if only by analogy, with the Marxist historicist theory of historical materialism. This latter regards the history of exploitative class relations as a propeller of the development of productive forces – a process which creates the material possibility for class relations to be transformed into social relations of equality. We see in Bookchin's perspective a similar sense of social relations as a promoter of the development of productive forces and thus justified as a means of creating conditions for individual liberation as asocial self-sufficiency.

THEORY OF SCARCITY AND THE SENSE OF TECHNOLOGY

In reconstructing Bookchin's theory of scarcity, we propose the hypothesis that while Bookchin regards scarcity as having many forms, he proposes that it has two different roots or sources: one presocial, the other social. This hypothesis is borne out in particular by his analysis of technology, as we see below, and casts additional light on his anthropological conception of human nature as asocially self-sufficient for scarcity, in the Bookchin perspective, is the antithesis of individual self-sufficiency. It is a condition of want which both activates and opposes the struggle for radical self-sufficiency. Before we examine the sense of this dialectic, we must look at that which constitutes a presocial condition of want. In other words, what precisely is the presocial condition of want? It is not merely a biologically determined want, but a lack, an insufficiency which prevents the individual from completing the chain from positing goals of self-sufficiency to developing and applying means to achieve them free of social mediation. It is a want as a lack of adequate tools and skills which prevents the individual from sustaining a life of self-sufficiency. It is of this which presocial scarcity comprises.

If scarcity occurs outside society – or precedes it in a sense – and if this scarcity denies the asocial individual an expression of his essence which is material and technical abundance to lead a life of independence and self-sufficiency, then some extra asocial instance is required as means to overcome this condition of presocial want. This instance is the social instance. The paradox with which Bookchin is confronted in his treatment of scarcity is that while human essence is asocial independence, it cannot be achieved outside society if such a 'state of nature' is marked by scarcity – and it is necessarily marked by technical scarcity for solitary man cannot alone develop the necessary technical expertise to overcome it. It is thus that structured social relations emerge.

If this interpretation is correct, we must quality our first contention regarding the ontological status of social relations. We can say that in Bookchin's construction social relations are not only unnecessary but, indeed, deny the expression of human essence as asocially self-sufficient. On the other hand, however, they are a necessary medium which creates conditions in which man can achieve his essence. These conditions involve the negation of the social. In other words, the social is necessary only in its mediation between presocial scarcity and post-social liberation. It is an ontological necessity only in its mediating form.

How do social relations vanquish presocial scarcity? Not without hurdles and setbacks it would seem, for, in social relations is generated another form of scarcity – a social scarcity which is always ideologically mediated and justifies a disproportionate social distribution of resources and privileges. This is what Bookchin means when he contends that socially generated scarcity is one of artificial wants which have little semblance to authentic needs deriving from human essence as asocially self-sufficient. But social relations, even in their fabrication of non-authentic needs and even in their negation of individual freedom are, for Bookchin it would seem, a necessary evil in that they mediate the development of productive forces (technology) as a condition for individual liberation. Once, however, productive forces have evolved to anywhere near the present level, social relations become wholly unjustified. We see this in Bookchin's reading of the contradiction of contemporary society as having, on the one hand, created objective material conditions for the achievement of individual self-sufficiency, but on the other hand, denying the individual such freedom.

> Social hierarchy is undeniably real today in the sense that it stems from a clash of *objectively* conflicting interests, a clash that up to now has been validated by *unavoidable material scarcity*.[2] (First emphasis is that of the author: the second is my own).

Bookchin's assessment of modern technology as having a liberating capacity comes out in the comparison he draws between a libertarian society, i.e. his eco-utopia, and the Hellenic polis, Athens, the rare historical experience nurturing self-development of the citizen. The former is judged superior because it has the advantages of drawing from modern technology.

> Modern technology – 'hard', 'soft', 'appropriate' or as I would prefer to call it, liberatory – has finally made it possible for us to eliminate the fears which stalked Aristotle: 'an overpopulous polis/of/foreigners and metics/who/will readily acquire the rights of citizens. . .' To these potential upstarts, one might also add slaves and women. The leisure of schole – the freedom from labour – that made it possible for Athenian citizens to devote their time to public life is no longer a birthright conferred by slavery on an ethnic elite but one conferred by technology on humanity as a whole. That we may feel free to reject that birthright for a 'simpler', 'labour intensive' way of life is historic privilege that itself is conferred by the very existence of technology.[3]

Bookchin makes an even stronger defence of modern technology as a liberating force in his analysis of the counter-cultural movement of the late '60s and early '70s which he sees as being initiated by middle class university educated youths for whom the principles of capitalism appeared to be outmoded by modern technology. Here, in this socio-historical context, according to Bookchin, liberation is possible in part because of the present developed state of modern technology.

> The work ethic, the moral authority imputed to material denial, parsimony, and sensual renunciation, the high social valuation placed on competition and 'free enterprise', the emphasis on a privatization and individuation based on egotism, seemed obsolete in the light of technological achievements that offered alternatives entirely contrary to the prevailing human condition – a lifetime free from toil and a materially secure social disposition oriented toward community and the full expression of individual human powers. The new alternatives opened by technological advances made the cherished values of the past seem not only obsolete and unjust but grotesque . . . there is no paradox in the fact that the weakest link in the old society turned out to be that very stratum which enjoyed the real privilege of rejecting false privilege.[4]

Bookchin's whole discussion of 'human needs' confirms this reconstruction of his theory of scarcity. In his discussion, needs are conceived as being authentic if and only if they are those of the free individual and satisfied by his own devices without intervention or interference from others. It would stand to reason, then, that socially produced and socially satisfied needs depart from the needs the individual posits and satisfies in immediate relation with nature, for the authenticity of a need is dependent on the individual's ability to satisfy it. The critical point here is that the means must be asocial. If the need requires social mediation, it is satisfied in a structure of domination and the need is false. In that case, the need is socially generated and antithetical to human nature as asocially self-sufficient. In Bookchin's claim

> it is not true in the diminution or expansion of needs that the true history of needs is written. Rather, it is in the *selection* of needs as a function of the free and spontaneous development of the subject that needs become qualitative and rational. Needs are inseparable from the subjectivity of the 'needer' and the context in which his or her personality is formed.[5] (author's emphasis)

Needs are thus identified with or linked to the free choice of the individual which is determined by technical capability. The individual has free choice to the extent that his technical capability allows him.

But what is the basis of this technical capability? Or, in other words, what is the realm in which it operates and, more importantly, in which it is determined? It could not be the social realm; it could only be the realm of pure organic causality. That which governs technical capability, in the Bookchin schema, is the system of organic natural laws. The inherent logic of these laws alone determines the success or failure of realising posited goals since the means of satisfying such goals are asocial. In other words, the individual's freedom of choice in determining his own need and in satisfying it is governed by a knowledge of natural processes. It is such a knowledge which extends or enlarges the scope of individual self-sufficiency. It is a knowledge which translates into a technical capability. Insufficient knowledge of natural processes constrains and limits individual choice. Freedom and liberation of the individual involve not only severance of social relations but a liberation from ignorance of technical means as the only possible means, in this theory, of satisfying posited needs. In the logic of Bookchin's conceptualisation thus, the conditions for liberation from ignorance is what socially developed technology provides. The way Bookchin assesses a hunter-gatherer society against his libertarian ideal is very revealing of this point. He sees such a society as being free of social domination but nevertheless not liberated for it does not satisfy the knowledge condition of freedom of choice of the individual owing to the low level of development of technical knowledge.

> Although a hunter-gatherer community may be free from the needs that beleaguer us, it must still answer to very strict material imperatives. Such freedom as it has is the product not of choice but of limited means of life. What makes it 'free' are the very *limitations* of its tool-kit, not an expansive knowledge of the material world.[6] (author's emphasis).

What is significant thus about Bookchin's assessment of technology is what it reveals about the mode of reproduction of the individual. It is not a social mode of reproduction but rather an asocial one. It is a technical activity by which the individual, in an immediate relation with nature, reproduces the conditions which sustain his asocial independence.

While Bookchin's analysis of technology as containing a liberating capacity in the transition from hierarchy to eco-utopia provides further evidence of the tenability of our interpretive hypothesis, it poses a fundamental contradiction for the problematic of his socio-ecological theory. This problematic as we noted earlier, poses a relation between man and nature not as one of control or domination of nature but rather in terms of a relation of cooperation with nature where nature itself acquires a subjective constitution. This relation is posited as one between two collaborative, communicating subjects, man and nature. If his assessment of technology is to be consistent with his problematic, this technology would have to comprise a relation of cooperation between human being, i.e. producer, and organic nature as two equal entities endowed with reason or subjectivity. This technology to which Bookchin appeals, however, is constituted within a relation which, if characterised as cooperative, is not a form of cooperation between two subjects but rather as a relation in which nature is subjected to a series of manipulations as a means of determining cause/effect relations. Necessary and sufficient causal conditions for a desired effect are not combined on the basis of communicating to nature an intersubjectively understood message but rather on the basis of controlled operations and experimentation in which the linkage of cause and effect is achieved, often by the process of elimination of error.[7]

To be consistent with his problematic Bookchin would have had to denounce the technology to which he refers as being one of domination of nature and proposed a different technology comprising a radically transformed relation between the human being and nature as an intersubjective relation. In failing to do this and in proceeding rather with an appeal to presently constituted western technology as a determinant factor for changing domination and hierarchy to equality and harmony, is Bookchin simply committing an error of judgement? That is, does his error lie in his calling upon a technology emerging from an instrumental relation between man and nature where he could and should have called upon a new technology – an intersubjectively constituted one? The question could only be posed thus if there exists the logical possibility of a reconstituted technology within a relation of intersubjectivity between the individual as producer and organic nature. It is the question which Habermas[8] puts to Marcuse in a critique of the latter which is resolved in a theoretical demonstration of the relation between producer and organic nature as being nothing other than instrumental. It is in this context that Habermas advances his notion of the logical structure of work as necessarily an instrumentally constituted relation between producer and organic nature. If Habermas's arguments regarding the logical structure of work as an instrumental relation between producer and nature are valid, Bookchin could not have called upon a new technology of intersubjectivity between producer and nature on logical grounds. The contradiction between his analysis of technology and his socio-ecological problematic remains thus necessarily unresolved. This contradiction is heavy with implication of an asocially self-sufficient anthropology as underlying his socio-ecological theory and indeed his very problematic.

Let us return, for a moment, to Bookchin's claim of the possibility of liberation in view of the present stage of development of technology. The argument is one which holds that knowledge of the laws of organic causality as cause/effect relations of organic processes are potentially liberating in the change from hierarchy to freedom. What form does this freedom necessarily take? It can only be a freedom deriving from the knowledge of the laws of organic causality and applied in combining necessary conditions as cause to produce a desired effect. However much this knowledge has been developed in social structures of hierarchy, Bookchin claims that it can now be put to use by the individual to assure his health care, his nutrition needs, his lodging requirements without recourse to alter as a social instance. The individual becomes self-sufficient with this knowledge by being liberated from subjecting to another's will. It is a radical self-sufficiency of the individual as solitary individual who is able to achieve the goals he posits without recourse to others. But if this self-sufficiency as freedom is one which necessarily subjects nature to a control in a process of combining and linking causal conditions to desired effects, nature remains a structure of organic causality however much it is posited, at the level of the theoretical problematic, as being subjectively constituted. Bookchin's individual departs from the structure of social hierarchy to enter the structure of organic causality. Bookchin's problematic, however, disallows the conception of freedom as a reversal to a state of nature where organic causality prevails and where harmony is the harmony of organic nature. It does so by interpellating nature as subject and positing the relation between man and nature as a social relation where all structured social relations necessarily hierarchical in Bookchin's schema are severed.

POLITICAL THEORY OF ECO-UTOPIA

Our hypothesis that an asocially self-sufficient anthropology underlies Bookchin's socio-ecological theory casts some light on the sense of 'libertarian society' or 'eco-utopia'. For Bookchin, eco-utopia marks an end to social hierarchy and the advent of a state of harmony in interpersonal relations between individuals and between individual and organic nature. That which converts social hierarchy to eco-utopia is self-activity – an act comprising simultaneously a refusal to comply to the will of another, i.e. externally defined direction, and the pursuit of self-defined goals. The change from social hierarchy to eco-utopia is achieved by the transformation of the individual into subject, a transformation mediated by self-activity. The key to understanding this eco-utopia thus is self-activity. Indeed, self-activity is the political theory of revolution from hierarchy to utopia and a political theory of utopia itself.

Hierarchical stratification, for Bookchin, is both a social and technical division of labour which denies the individual the possibility to be subject in the sense of defining his own needs and goals and commanding the means to achieve them. It is a division between, for example, workers and employers with the concomitant disproportionate control of means of production, laymen and experts with the disproportionate command of knowledge necessary to attend to one's own needs directly. In this social and technical division of labour is rooted authority which, as we saw, is a relation of command and obedience owing to a concentration either of wealth, power and privilege, knowledge, where one commands by virtue of his wealth, his prestige, his power, and his knowledge, and the other subjects to this command owing to his subordinate position. In this social structure decision-making is concentrated such that some decide, others comply. The effects of these decisions, therefore, have a social scope. They are decisions which involve social compliance. Self-activity takes the individual out of this social and technical division of labour by concentrating decision-making at the level of the individual and more importantly by limiting the scope of the effects of these decisions to the individual, himself. This is what Bookchin emphasises in defining the individual as his own centre of decision-making. He attends to these decisions and fulfils them himself. 'The power of man over man can be destroyed only by the very process in which man acquires power over his own life.'[9]

It is clear that this self-activity can be nothing other than self-sufficiency. It is the activity by which the individual achieves and assures the conditions of his own reproduction directly. This self-activity involves a knowledge capability necessary to realise the self-defined goals directly without recourse to the expertise of alter. It involves achieving this knowledge in a process of self-development, a knowledge which as we saw in the earlier section is a knowledge of organic causality which, with one stroke, negates the social division of labour and the technical division of labour by providing the individual all the holistic knowledge of the various facets of the total operations. Both become unncessary as the individual becomes self-sufficient.

The politics of social change thus is the mobilisation of individualised activity which strives to achieve radical self-sufficiency. This is Bookchin's meaning of the above-cited proposition that, 'the power of man over man can be destroyed only by the very process in which man acquires power over his own life and in which he not only 'discovers' himself but, more meaningfully, in which he formulates his self-hood'.[10] Bookchin, accordingly, rejects categorically all historical socialist revolutions including the Soviet, Chinese and Cuban for their entrenchment of relations of domination between leaders (vanguard political party) and led (masses). Liberal democracy is rejected as well, principally on the grounds of its having an essentially hierarchical structure. Insofar as it has evolved into electoral politics and indirect democracy, it constitutes a division between the governors and governed, the elected and electors in which decisions are concentrated on the side of the governors. It merely reproduces, in a political form, hierarchical relations of command and obedience.

Bookchin is more attracted to the 'affinity group' of the Iberian Anarchist Federation of pre-Franco Spain and to the American counter-culture movement of the '60s and early '70s for their effort to create conditions in which the individual can manage his own life. He describes how affinity groups,

> could easily be regarded as new type of extended family, in which kinship ties are replaced by deeply empathetic human relationships – relationships nourished by common revolutionary ideas and practice. Long before the word 'tribe' gained popularity in the American counter culture, the Spanish anarchists called their congresses *assambleas de las tribus* – assemblies of the tribes. Each affinity group deliberately kept small to allow for the

> greatest degree of intimacy between those who compose it. Autonomous, communal and directly democratic, the group combines revolutionary theory with revolutionary lifestyle in its everyday behaviour. It creates a free space in which revolutionaries can remake themselves individually, and also as social beings.[11]

And of the counter-culture, he argues that,

> young people were quite right in sensing that existential personal goals must be defined and striven for even today, within the realm of unfreedom, if future revolutionary changes are to be sweeping enough and not bog down in bureaucratic modes of social management.[12]

A more distant historical experience attractive to Bookchin is the Hellenic polis which he regards as one encouraging individual citizen participation in public life. If our interpretation of Bookchin, so far, is correct, his polis or state is not an emulation of the Hellenic polis. It is rather a sum of individual decision makers on matters of radical individual self-sufficiency including the conditions of its maintenance. We find in this state a common interest of individuals but such an interest could not be likened to a Rousseaunian general will nor to a Marxist community. Indeed, there is no social interest here at all, but rather an asocial interest. This common interest can be seen as being free of internal contradiction in contrast, for example, to the bourgeois individual interest of private property in relation to the labouring class. The absence of internal contradiction in the common interest of Bookchin's individuals owes to its being an asocial structure. It is the harmony of a state of nature. This, we suggest, is the form of state or eco-utopia in Bookchin's theory. It has also evident implications for his theory of ecology.

ECOLOGY AND UTOPIA

We now see the sense of ecology in Bookchin's eco-anarchism. If our hypothesis that Bookchin's socio-ecological theory is marked by the premise of an asocially self-sufficient anthropology is correct, his theory can allow only organic causality as the sole source of determination. Self-sufficiency is the effect of the correct combination of necessary and sufficient causal conditions. This casts light on the problematic of this socio-ecological theory. As one which poses the relation between producer and nature as an intersubjectively constituted relation, nature is interpellated as subject in the same way in which the individual is interpellated as subject and as one believing himself to be the source of determination in unison with nature. Leaving aside, for the moment, the question of the character of this reconstitution of the individual and nature as subjects, it is a problematic which brings ecology to the fore. Indeed, in the context of this problematic Bookchin is able to attribute a primacy to his subjectively constituted nature. It is thus that he can depict 'wastelands' as the effect not of society's manipulation of nature but of nature's resistance to manipulation as an instance of the assertion of a higher power, force, and reason.

> No historic examples compare in weight and scope with the effects of man's despoliation – and nature's revenge – since the days of the Industrial Revolution, and especially since the end of the Second World War. Ancient examples of human parasitism were essentially local in scope; they were precisely *examples* of man's potential for destruction, and nothing more. (. . .) Modern man's despoliation of the environment is global in scope, like his imperialism. It is even extra-terrestrial, as witness the disturbances of the Van Allen Belt a few years ago. Today human parasitism disrupts more than the atmosphere, climate, water resources, soil, flora and fauna of a region: it upsets virtually all the basic cycles of nature and threatens to undermine the stability of the environment on a worldwide scale.[13]

His derivation of the model of eco-utopia from the properties he attributes to nature is instructive of this. From ecology Bookchin derives the principles of diversity and the principle, which we may call 'horizontality' or the absence of all hierarchy.

According to him, the ecological principle of diversity contends that ecological stability derives from diversity. The more varied the species of fauna and flora constituting a given eco-system, the more stable that system is in terms of its cycle of reproduction as a total eco-system. This means that any process of homogenisation of the eco-system threatens the capacity of the system to reproduce itself. Bookchin bases the following observation on the work of Lotka, Volterra and Bause:

> The greater the variety of prey and predators, the more stable the population; the more diversified the environment in terms of flora and fauna, the less likely there is to be ecological instability.

> Stability is a function of variety and diversity: if the environment is simplified and the variety of animal and plant species is reduced, fluctuations in population become marked and tend to get out of control. They tend to reach pest proportions.[14]

He concludes, thus, that much of the ecological crisis produced by the Industrial Revolution is linked to the increasing reduction of diversity in nature as various regions become specialised in the production of a given commodity such as coal or other minerals, lumber, etc.

His principle of diversity in nature becomes an important principle of eco-utopia in two senses. It becomes a principle of social diversification and differentiation within an association of individuals, which he refers to as community, and between communities. The free expression of differences in individual subjectivity, according to him, rather than reflecting social divisions between groups and individuals of disproportionate power, reflect the unrepressed spontaneity of individual reason. It also becomes a principle for the diversification of production. Indeed, this principle of diversity underlies the whole notion of ecotechnology as a means of production. Bookchin defines it as one which incorporates a diversity of means of production compatible to natural, ecological diversity. This includes, for example, the diversification of sources of energy including solar, wind and water energy. A selection of any single one of these would be as 'regressive as adopting nuclear energy', argues Bookchin. Agricultural production which emphasises diversity is rational both as a sound ecological principle and a principle of self-sufficiency. This self-sufficiency is conceived both in terms of satisfying various needs and in terms of constituting the conditions of production in which individuals are active, self-sufficient subjects rather than dependent on others. About this sense of diversity and variety as ecotechnology, Bookchin writes,

> One can well imagine what a new sense of humanness this variety and human scale would yield – a new sense of self, of individuality, and of community. Instruments of production would cease to be instruments of domination and social antagonism: they would be transformed into instruments of liberation and social harmonization. The means by which we acquire the most fundamental necessities of life would cease to be an awesome engineering mystery that invites legends of the unearthly to compensate for our lack of control over technology and society. They would be

> restored to the everyday world of the familiar, of the *oikos*, like the traditional tools of the craftsman. Selfhood would be redefined in new dimensions of self-activity, self-management and self-realization because the technical apparatus so essential to the perpetuation of life – and today, so instrumental in its destruction – would form a comprehensible arena in which people could directly manage society. The self would find a new material and existential expression in productive as well as social activity.[15]

In diversity there is also harmony, Bookchin claims by pointing to the homogenisation and discord of the structure of modern cities as one confirmation of the validity of the ecological principle of diversity.

The second principle of ecology which Bookchin proposes is what we have referred to as horizontality. According to Bookchin, it is the principle which captures the sense of the essential structure of nature, namely non-hierarchy. Accordingly, he depicts nature as being a varied diversity of elements which combine to form an integral, organic whole. No element, in itself, is superior to another or stands above it in a relation of authority. The diversity in nature in no way translates into hierarchy of elements but rather into a differentiated organic totality of unique components. This individual uniqueness does not confer on any given part privileges denied to others or power to control others. Indeed, Bookchin insists that 'there are no hierarchies in nature other than those imposed by hierarchical modes of human thought, but rather differences merely in function between and within living things'.[16]

From his second principle of the dynamics of nature, he derives further features of eco-utopia. Equality is principal amongst them which he takes to be 'in the nature of things'. Indeed, any subordinate/superordinate relation of authority is a deviation from the natural order, according to him. It does violence to the innate nature of man which is to be independent and subject to his own will and consciousness in a direct relation with nature. This is the reason and will of a free individual as one that is neither subject over another nor to another. Bookchin derives his differentiation as a uniqueness of subjectivity between individuals from his differentiation of elements of the natural order. It is thus that differentiation between individuals becomes not a specialisation of function, a division of labour but merely each individual's 'doing his own thing'. The absence of hierarchy in diversity, according to him, contributes to stability and harmony in nature. Similarly, the absence of social stratification either as social or technical division of labour makes not only for self-sufficient, independent individuals but for stability and harmony in eco-utopia. For Bookchin, present society has nothing

> that would seem to warrant a molecule of solidarity. What solidarity we do find exists despite the society, against all its realities, as an unending struggle between the innate decency of man and the innate indecency of society. Can we imagine how man would behave if this decency could find full release, if society earned the respect, even the love, of the individual? We are still the offspring of a violent blood-soaked, ignoble history – the end product of man's domination of man.[17]

THE SOCIAL BASIS OF THE ANTHROPOLOGY OF ASOCIAL SELF-SUFFICIENCY

What sense can be made of the principles which Bookchin attributes to nature? Are the principles of diversity and of non-hierarchy or 'horizontality' immediate descriptions of the manifestations of nature? An answer in the affirmative would imply that sense perception is an immediate registry of the characteristics of the perceived in a way that the perceived object conveys directly and immediately its characteristics to the perceiver. This would be a form of empiricism which would be difficult to defend. If on the other hand, these principles are taken to be other than mere descriptions of the manifestations of nature, a further question arises as to what intervenes between the perceiver and the object perceived. In other words, if these principles are regarded as being reconstituted conceptions of the characteristics of nature as a concrete object, the question which arises is what considerations, assumptions, notions, etc., enter into the theoretical reconstitution of the concrete object?

Can we say that this conception of nature as free of hierarchy is a scientifically reconstituted conception and that it provides the accurate and correct understanding of how nature is comprised, or, more in the spirit of the notion, of how nature comprises itself? Or do we say, on the other hand, that this conception of nature is an ideological conception, i.e. an ideologically constituted conception of nature? There seems to be very little basis on which to claim that the notion of a subjectivised nature is anything other than a case of anthropomorphisation of organic causality such that it acquires anthropological characteristics. But this last proposition that a 'subjectivised nature' is an ideologically constituted notion raises the fundamental question, namely, the basis of determination of such an ideological conception.

Earlier, our review of Marx's analysis of determination of classical liberal notions such as individual subjectivity, equality, freedom, etc., helped to situate the social structures of which principal notions of neo-Malthusianism and expansionism are the effect. We identified the particular interest structure of these specific conceptions as capitalist class dominance. Can the theory of ideology which Marx developed in *Capital* and which served to interpret neo-Malthusianism and expansionism, help to elucidate eco-anarchism in a similar way? Is it possible, following Marx, to identify a particular interest structure in eco-anarchism?

If Marx's analysis reveals that individual subjectivity, equality, freedom and various other notions of classical liberalism comprise a bourgeois ideological structure, it does so at the abstract level of mode of production with two structurally determined classes: the capitalist class and the working class. But his construction of mode of production does not accommodate or allow an interpretation according to which the notions of individual self-sufficiency, 'nature-subject' and other principal notions of eco-anarchism are an ideological effect of the relations of domination and exploitation from the side of the working class, i.e. the principles of ideology of the working class. If this is correct, it would imply one of two things: (a) that Marx was wrong in the theoretical propositions he advances in *Capital* regarding the capitalist mode of production and thus conceptions, ideas, beliefs are not socially determined as class conceptions, ideas, or beliefs. (This is the conclusion most compatible with Bookchin's own claims in regards particularly the anthropological assumptions of human nature.) Or (b) that Marx was right in his analysis but that these particular notions as socially determined notions and indeed socially determined as specific class notions, cannot be linked either to the capitalist class or the working class as class. In order to establish the specific social determination of these notions as class notions we must move from the abstract level of mode of production to the concrete level of social formation, the locus at which modes of production are actually reproduced. It is the latter possibility that can be more tenably retained. Indeed, as social formations are historically determined articulation of modes of production in which the dominant mode is determinant, classes in addition to the two of the dominant mode combine in a complex articulation of class struggle. In such historically determined social formations as the United States, Canada or those of Western Europe two factors can be noted in the articulation of modes of production. The first is that dominant in the articulation is not only the capitalist mode of production but indeed the monopoly form of this mode. The second is that constituent of these social formations is the petit bourgeois class as well as other classes, including but not exclusively, the bourgeois class and the working class.

The thesis we should like to advance here is that the principal notions of eco-anarchism are class-determined notions and more specifically they are petit bourgeois class notions. In making this claim we invite, perhaps, a heavy burden of proof, for if it is conceded that ideas, beliefs, values are socially determined and conceded also that this process of social determination of ideas, values, etc. in a capitalist social formation is a process mediated by a class structure and thus, ideas, values, etc. are effects of these class relations and class struggle, it is unclear, at first sight at least, how the radical ideas and values of eco-anarchism can be construed as being petit bourgeois. It is also unclear how the conceptions and principles expounded by a social critique such as that of Bookchin can be considered to express the values and aspirations of the petit bourgeoisie as a class.

First of all, let us dispel the simplistic idea that the sense of the relationship between Bookchin's theory and petit bourgeois ideology is a mere designation as spokesman, either by 'class' election or self-appointment, of Bookchin to express the aspirations of this class. This is not the sense of social determination nor the sense in which methodologically we may refer to a corpus of coherent thought such as that of Hobbes or Locke as being bourgeois.[18]

Since social determination is a material process we begin with this, and find that it is this very process which leads us to the identification of the specific class character of a given corpus of thought. An examination of the place of the petit bourgeoisie in this material process will elucidate the sense of this proposition.

The dominant class relation in capitalist social formations is the relation between the bourgeois class and the working class and it is this dominant relation which determines the place of the petite bourgeoisie in the class struggle. Let us examine this proposition more closely. In its economic form, or, in other words, on the level of economic relations of production, the relation between the bourgeois class and the working class is a relation of exploitation in the specific sense of production of surplus-value (by the working class) and its appropriation (by the bourgeois class) owing to the relation of these classes to the means of production i.e. capitalist ownership and control of means of production, working class dispossession of the means of production. But this economic relation as a two-class relation of polarity between the bourgeoisie and the working class determines also the specificity of the other classes by exclusion. That is, the very dominant economic relation which constitutes two opposing conflictual classes, also denotes non-bourgeois and non-working classes by a negative criterion of exclusion. The middle classes are 'negatively' denoted as classes whose relation to the means of production is neither capitalist (i.e. private appropriation of surplus-value) nor working class (i.e. direct production of surplus-value) but rather ones which, as in the case of the old middle class of independent direct producers (artisans, family production), own their means of production without appropriating surplus value and ones which, as in the case of the new middle class, are salaried employees without being directly productive of surplus-value.

The economic relation reveals by exclusion not one but two classes – the old and the new petit bourgeois classes. Nicos Poulantzas was amongst the first[19] to point out this 'negative' structural determination of the old and the new middle classes as two distinct economic classes in exclusion from the bourgeois class and the working class. Poulantzas takes this negative criterion which sets the old and new petit bourgeois classes apart from the bourgeois and the working class not as expression of classlessness nor even of class disparity in a theoretical sense of two classes but rather of one, single class whose structurally determined unity is confirmed on the ideological and political levels.

As a class distinct from the bourgeois class and the working class the petite bourgeoisie is not necessarily unified in the ideological position or in the political position it adopts at any given time. Nor, in this internal division, are the position or positions it adopts ones necessarily different from and conflictual to those of either the bourgeois class or the working class. The petite bourgeoisie may, in the ideological and political class struggle, by the very position or positions it adopts, 'shift' totally either to the bourgeois class or the working class or divide such that a part aligns with the bourgeoisie, a part with the working class and thus deny to itself an independent political and ideological position without negating itself as a distinct, structurally determined class in the capitalist social formation. Without entering into the question, important and complex as it is, of the theoretical relation between structure and juncture in class struggle, suffice it to say that one is not independent of the other as two separate arenas in which the class struggle unfolds but rather as two dialectically related determinants of class struggle where one may be the source of change or containment of the other in the very class struggle.

But a further look is necessary at the sense in which the petite bourgeoisie is a structurally determined distinct class in ideological and political relations in capitalist social formations, before anything can be said of a petit bourgeois ideology, the very matter which interests us here. In fact the structurally determined ideological and political relations of the petite bourgeoisie as a distinct class, separate and distinct from the bourgeoisie and the working class, is an integral part of the reproduction of the capitalist social division of labour – indeed, of capitalist relations of production. For in the division of labour between productive and non-productive labour as an economic relation, there are also ideological and political elements which serve to distinguish and separate these two forms of labour into distinct classes. The ideological and political elements, which we shall examine below, dictate against, as Poulantzas cautions, our considering the division between productive and non-productive labour as a mere technical division of labour.

The relation between the petite bourgeoisie and the working class in the social division of labour is one of petit bourgeois surveillance and supervision of productive labour. It is a political relation of domination as a superordinate/subordinate relation in which the social division of labour dominates the technical division of labour. The work of supervision is a political relation of domination in the very process of production as it involves the exercise of power in the extraction of surplus-value from the workers. But this power derives from the dominant position of capital and the capitalist class and not from the position of the petit bourgeois class in the relations of production. The petite bourgeoisie exercises capitalist power without itself being part of the capitalist class. For the relation between the petite bourgeoisie and the capitalist class is still a relation of exploitation, in that part of the work of non-productive labour remains unpaid as surplus labour without its being surplus-value as in the case of productive labour.

The relation of the petite bourgeoisie and the working class is a relation of domination in an ideological sense as well. The intellectual labour of the petite bourgeoisie is dominant in relation to the manual labour of the working class in the capitalist productive process. This domination is the effect of the mode of development of productive forces in the capitalist relations of production. Science and technology are the critical factors here. Science consistently claims neutrality from class values or class interests, which neutrality is essentially a denial of any effect of social relations of production on its conduct of inquiry and on its theoretical product, i.e. scientific theory. The technological application of science, however, reveals the claim to neutrality to be ideological in that the purpose of this technology is the increase of the rate of relative surplus-value, i.e. the increase of the rate of exploitation.[20] To clarify the ideological sense of this relation of domination between intellectual and manual labour and to demonstrate how this relation is a vehicle of capitalist class dominance, let us follow Poulantzas in his analysis of the labour of engineers and technicians as illustrative examples of this relation.

Along with Poulantzas, we must dispel the claim that a quality of superiority inheres in the technical nature of intellectual labour as opposed to manual labour. Intellectual labour becomes qualitatively superior in capitalist relations such that the division of intellectual labour as intellectual and manual labour as manual is an ideologically constituted division of labour. As a relation of domination-subordination, it is determined not by the separation of science from the direct producers but by the separation of the direct producers from the means of production of which the very separation of science from direct producers is partially an effect, and more importantly, it is an effect which determines the ideological relation of dominance of intellectual labour over manual labour.

In what sense does the labour of technicians and engineers exemplify the ideological and political aspects of dominance as the structurally determined dominant aspect of class determination of the petite bourgeoisie in its relation to the working class? It is particularly exemplary in that this labour is both intellectual and productive of surplus-value and it can display the relation in its very dual constitution. While the labour of technicians and engineers tends to be productive labour, it is the intellectual labour aspect which is dominant as technical expertise and as such ideologically constituted as dominant in relation to productive labour. The distinction serves ideologically to qualify intellectual labour as superior to manual labour and thus to set the petite bourgeoisie apart in mannerism, attitudes, speech, knowledge, as various expressions of privilege of intellectual labour as distinct from manual labour. This is so of a whole range of intellectual labour, widely disparate in regards to any other criteria, from scientists to sales persons to clerks to service persons. This very ideological distinction is reproduced in and by the school system which serves to segregate intellectual work from manual work by channelling the petite bourgeoisie to the intellectual road and the working class to the manual road in the social division of labour.

The social division of labour between intellectual and manual labour as a relation of political and ideological domination-subordination acquires a full materiality in bureaucracy. In addition to reflecting the separation of intellectual from manual labour, bureaucracy is the locus of articulation of ideological and political relations of domination of the petit bourgeois class over the working class. This relation of dominance of the petit bourgeois class over the working class is, itself, of course, determined by the relation of dominance of the capitalist class over the working class. This capitalist-determined relation of domination of the petite bourgeoisie operates by according the petite bourgeoisie not its own independent class power but the exercise of power of the capitalist class. In this way the petite bourgeoisie becomes integral to the legitimation of capitalist power over the working class. Here lies the contradiction of the class position of the petit bourgeois class in capitalist relations: it both exercises power and is powerless in terms of its own independent class power. Let us examine, if only briefly, some aspects of bureaucracy to situate the full sense of this contradiction.

Bureaucratisation is an organized system of 'rational', impersonalised distribution of activity and competence of the state apparatus and of private corporations including banks, financial institutions and industrial enterprises. In this system of impersonalisation of function, of bureaucratic secrecy and of monopoly of knowledge, prevail relations of authority such that agents relate 'authoritatively' not only to each other according to defined rules of command and obedience but regard these relations as legitimate. What is critical here is that in these ideological and political relations of domination-subordination within the petite bourgeoisie itself, as well as between the petite bourgeoisie and working class, operates the domination of the bourgeoisie not only over the working class but also over the petite bourgeoisie. For the authority exercised by this class in its relations with the working class is not power as bourgeois class power but rather authority in that not only the working class but the petite bourgeoisie itself is subject to bourgeois class domination. This contradiction in the very structural constitution of ideological and political relations as being both on the side of domination (in relations of authority) and on the side of dominated (as subject to bourgeois class power) conditions the ideology and politics of this class. It is rarely a position reflecting a unified class around which this class gravitates as a class. More often it is a divided class adopting positions akin to that of bourgeois class or to that of the working class, one fraction or fractions shifting to the former, the other to the latter without, however, embracing either in purely its bourgeois or working class form but rather in a petit bourgeois mediated form as we shall see below.

The hierarchical distribution of authority, i.e. disproportionate distribution of authority, including impersonalisation and bureaucratic secrecy, also have an effect on the ideology and politics of this class. The immediate structural effect is one of 'isolation' of agents of the petite bourgeoisie, in Poulantzas's characterisation, and this has an effect of divisiveness of the class and of relative absence of class solidarity in politics and ideology.

The contradiction of this class in its ideological and political aspect is determined in the last instance by the economic relations of production which also determine the economic contradiction of this class. As we have noted earlier, this class is neither productive of surplus-value nor appropriator of surplus-value and hence distinct from both the working class and the capitalist class without however being autonomous and independent from the bourgeois-working class relation. It is a class 'separated' from the relation of private appropriation of surplus-value and a class dependent on this relation of capitalist exploitation, a factor which conditions its position in the economic class struggle as both one of alliance with the bourgeoisie in increasing relative surplus-value (intensifying the rate of exploitation of the working class) by the scientific-technical application of scientific research, and of alliance with the working class in demands for job security, salary increases and generally favourable conditions of work.

These economic, political and ideological contradictions account in large part for the conservatism as well as for the radicalism which the petite bourgeoisie, either united or divided, assumes in the ideological and political class struggle. The form of conservatism is not identical to bourgeois conservatism although it is consistent with bourgeois class dominance. Appeal to traditional values of family, church and state, assertion of institutional authority, appeal to individualism as equality of opportunity and upward mobility, and so on, is a specifically petit bourgeois form of conservatism. But it is in its political and ideological radicalism that there appears the contradictions of this class. It is not a proletarian radicalism of a working class revolution for the socialisation of means of production and thus of the radical transformation of the capitalist relations of exploitation and domination. It is a radicalism as opposition to the capitalist relations of exploitation and domination from a place separate from (though integral to) this dominant relation, for this relation, in its dominance, is a relation between the working class and the bourgeoisie. It is an anarchist radicalism. It is opposition to domination and exploitation from the perspective of a dominated and exploited class but whose form of domination and exploitation is dependent on the capitalist exploitation and domination of the working class.

This radicalism of the petit bourgeois class struggle for liberation from domination and exploitation assumes not a collective form of class solidarity (for as a class, it is barely well placed to achieve structural change), but a highly or extreme individualist form of action in which the struggle for liberation becomes an existentialist individual action for individual independence and individual autonomy. On the economic level, the struggle of the petite bourgeoisie assumes a form of independent production which in its radicalism is not immediately threatening to the relation of domination of capital. While the petit bourgeois class struggle in the economic class struggle is waged on two fronts – one by the old petite bourgeoisie, and other by the new petite bourgeoisie – each is contained by its structural contradiction and this is reflected in the impotent struggle for individual independence. The independent direct producers, even as owners of their means of production, believing themselves to be independent, control nothing of the social productive process and remain victim to the increasing economic rationalisation and capitalist absorption of family enterprises.[21] The petit bourgeois salaried workers, rendered superfluous in the continuingly accelerated process of mechanisation and automation, find no refuge in recycling or adult education for specialised expertise but rather become part of what is now being commonly referred to as 'the reserve army of intellectual workers'. There is, of course, as we shall discuss below in reference to eco-anarchism, the form of petit bourgeois economic radical action as development of means of individual reproduction in the various urban cooperatives some of which originate in the 'back to the land movement'. This is a form of radical action against the capitalist relations of domination and exploitation as rupture or withdrawal from this relation. It is individualised action even when whole groups participate because it is not action immediately affecting the capitalist relation of exploitation and domination – a relation between the capitalist and working classes. It is here that we can grasp the source and significance of the anthropology of self-sufficiency underlying Bookchin's theory as a form of petit bourgeois radicalism. But before we examine the specific relation between Bookchin and petit bourgeois radicalism, we must situate some additional aspects of this radicalism in its economic, political and ideological forms and of its structurally determined contradictions on the level of economic, political and ideological class struggle.

On the level of economic class struggle, radicalism as petit bourgeois struggle against capitalist exploitation can be described almost as 'passive resistance' even in its form of withdrawal from capitalist relations and this derives from the economic position of this class as separate from, though integral to, the dominant relation between capitalist class and working class. Petit bourgeois radicalism assumes a more visible, forceful form on the level of ideological class struggle and the dominance of this form of petit bourgeois radicalism is itself, as we saw, structurally conditioned. In its critique of capitalist domination, the relation of capitalist exploitation becomes severed from the relation of capitalist domination. This is well illustrated in Bookchin's separation of relations of domination from relations of exploitation which assumes that relations of domination have no necessary apparent relation of overdetermination with relations of exploitation. All forms of authority are seen here as channels of authority and repression.

The principal theme of radical ideology thus is liberation as individualist self-assertion. Individualism, as independent capitalist entrepreneurialism, is replaced by an individualism in a different guise. The 'individual' is replaced by 'self' and individual independence by self-liberation in which this 'self' severs from institutional relations and ties which dominate and tie him or her down, and defines goals and develops means to achieve them free from dependence or interference from others. Self-realisation and self-sufficiency are supreme goals. The community of self-defining and self-sufficient individuals replaces collectivism and socialism.

On the level of the political class struggle petit bourgeois radicalism assumes an extreme individualist form as resistance and opposition to political parties of any ideological colour, of electoral politics, and of indirect liberal democracy. The petit bourgeois political radicalism is one of direct, immediate action through citizen committees and neighbourhood councils for greater citizen 'non-state' control of urban housing, education, urban transportation and so on, and generally for the reduction and diminution of the sphere of control by the state apparatus. The structural contradiction of this class limits petit bourgeois ideological and political struggle for change of the capitalist structure to an individualised action where the struggle for transformation of the capitalist structure (capitalist/working class) remains removed from this relation. The individualised struggle reflects the contradiction and the limits of petit bourgeois class struggle even in its radical form.

These considerations cast considerable light on the eco-anarchism of Bookchin. It is a theory constituted from the perspective or situation of the petit bourgeois class. His theory of liberation bears many characteristics of structurally determined petit bourgeois relations of exploitation and domination to which this class submits in capitalist relations. That the critique of capitalist relations becomes a critique of relations of domination severed from relations of exploitation is no mere oversight on Bookchin's part. It is not an expression of confusion as some intellectual insufficiency in scientific theorising but rather of the structurally determined contradiction of the petit bourgeois class. It is the perspective of this class which conditions this particular understanding of the objective process of capitalist production. It is thus that Bookchin's separation of relations of domination from relations of exploitation becomes an anarchist critique of all forms of authority as channels of repression. It is thus that relations of exploitation have no necessary relation of overdetermination with relations of domination for him. In this severance we find a petit-bourgeois-class-determined answer to the theoretical question of basis of determination of authority or source of authority in Bookchin. The inevitable answer is social structure or social relation. Authority is conceived as being necessarily social and social as necessarily authoritarian.

This proposition determines the form which liberation from authority can take. It is not a change of structure. It is not a change of social relations but a break or severance from structure by means of self-sufficiency. It is not a change of the structure of domination to one of liberation, for domination inheres in structure and thus liberation can only be achieved with a break from structure.

If, of course, we read Bookchin on the surface, we find nothing of the kind. Eco-anarchist liberation as asociality would be quickly denied by any adherent to this theory, a denial which however, could not be sustained on the level of meta-theoretical critique. The real question thus is how can an asociality, which appears clearly a ludicrous assumption, form the grounds of a theory of social liberation? This assumption cannot be characterised in such terms as 'untenable', 'erroneous', 'arbitrary' or 'incoherent'. What has to be explained is how an assumption of asociality in all its untenability finds its way into theoretical construction. As we relate the theoretical construction to the class perspective of the petite bourgeoisie we see that not only is asociality structurally determined but so is its untenability itself.

As we discussed earlier, the petit bourgeois class is not constituent of the dominant capitalist relation between capital and productive labour. It is integral to this relation as a class apart from the two essential and opposite classes of the dominant capitalist relations. Change of the dominant capitalist structure as a structure of exploitation and domination cannot come from without this structure. It requires the involvement of the directly exploited and dominated class, i.e. working class. What prevents the incorporation of this working class involvement in working class terms is the perspective of the petit bourgeois class.[22] But the contradiction of this class in the class structure reflects the untenability of its solution for change, namely withdrawal from the dominant capitalist relations. It is thus that we can situate the anthropology of asocial self-sufficiency as a structurally determined petit bourgeois class form of radical anthropology.

Let us examine closer the structural position of the petit bourgeois class in the structure formative of the anthropology of asocial self-sufficiency – anthropology, which we argued, underlies Bookchin's theory. Bookchin, as we recall, rejects individual freedom as possessive individualism or as private entrepreneurialism. His liberating and liberated individual or 'self' is a self-defining, self-realising, self-sufficient individual. Such is, for Bookchin, the human subject expressing his essence as a 'self' who posits goals and develops such means to achieve them which do not involve submitting others to instrumental control, i.e. means for achieving his goals in a relation in which he is dominant and others dominated. Indeed, in depicting his liberated subject, Bookchin eliminates all instrumental relations which the subject, in pursuing his goals, may enter including his relation with organic nature. He does so by making organic nature itself a subject. In this way, the human subject, the 'self', secures the cooperation or agreement of nature in his activities. While depiction of the subject is explicit in Bookchin, asocial self-sufficiency is implicit but it is in terms of asocial self-sufficiency that the subject can be seen to be a solitary individual.

This solitary individual is the petit bourgeois subject which appears in eco-anarchism in the same way the bourgeois private individual appears in liberalism, namely as an anthropology. Just as in liberalism where the subject in the form of the private individual is posited not as an ideological subject, least of all a bourgeois ideological subject but rather as ontological comprising in itself epistemological truths, so in eco-anarchism, the self as the individual solitary individual is posited not as an ideological subject, least of all a petit bourgeois ideological subject, but an ontological one as the source of determination. The radical 'breaking out' of a structure is action free of structure. The petit bourgeois individualism in its anarchical petit bourgeois form is asocial action of self-sufficiency.

We come finally to the question with which we began this whole discussion of social determination of concepts, namely the notion of nature in Bookchin's theory. What considerations, assumptions, notions, etc., enter into the theoretical reconstitution of organic nature? We were sceptical of the view that Bookchin's depiction could emanate directly from nature as if organic nature converted itself into an object of knowledge in Bookchin's schema. We were more inclined to accept the view that the theoretical constitution of concrete objects into theoretical objects follows another course. It is the course of theoretical practice in which scientific theories and notions mediate the reconstitution of concrete objects into objects of knowledge. Without getting into the whole question of conditions in which theoretical practice yields scientific objects of knowledge, suffice it to say that scientific knowledge cannot be one into which ideological notions find their way disguised as scientific or ontological.

As regards the notion of nature in Bookchin's schema, what is disguised as ontological, particularly the characterisation of nature as being free of hierarchy, is an organic nature reconstituted in theory in terms of radical anarchist ideology. We see this ideology to be petit bourgeois. In its petit bourgeois form of radicalism it manifests the structurally determined contradiction of this class which limits radical class action to solitary individual self-sufficiency. The depiction of nature as diverse but egalitarian and free from hierarchy is thus a depiction which echoes the form of radical action of this class. It is a form which, in its limits, reflects the structurally determined contradictions of this class.

We see now the full sense of ecology in eco-anarchism. It is more than environmental preservation and conservation of nature. It is asociality. Nature as asociality is eco-anarchism or eco-anarchism is nature as asociality. The breaking away from social structure and the liberation of nature is the ecological theory of eco-anarchism, a petit bourgeois form of radicalism.

5 Ecology According to Gorz

The theoretical work of André Gorz has a narrower focus than does that of Bookchin. Its object is contemporary social relations in advanced industrial social formations and in particular capitalist social formations. In theorising his object Gorz appeals to Marx but the socio-ecological theory he arrives at has more in common with Bookchin than with any tenable reading of Marx. The key to this paradox is the implicit premise of Gorz's general social theory which is one and the same with that of Bookchin, namely an anthropology of self-sufficiency. It is this premise which clouds Gorz's reading of Marx and yields a theory of revolution as one of individual self-management which converges with Bookchin's theory of revolution as self-activity.

What light can the question of the nature of the link between Marx and Gorz shed on our interpretation of the principal themes of Gorz's general social theory in terms of this premise? Gorz builds his entire analysis on the assumption that activity is the medium by which human reason expresses itself. It is an act of production as a moment of synthesis of reason and substance. In other words, activity is the process of achieving rationally posited objectives. This activity is labour as a unity of reason and matter. So constituted, labour is the essential quality of the human being through which he expresses his creativity.

This notion of labour as human essence is the fundamental assumption of humanist Marxist thought.[1] It is clearly evident in many of Marx's writings, in particular *The Economic and Philosophic Manuscripts of 1844* and the *Grundrisse*. This is the notion which underlies the humanist theory of history as a progressive succession of forms of social labour propelled by a *telos* of free subjectivity of community, this free subjectivity being the human essence. Social labour captures the historicist Marxist sense of mode of production as a dialectical unity of forces and relations of production in which productive relations have primacy over productive forces and in which there is a sense of history as a teleological unity of modes of production.

If the notion of labour as 'human essence' is the link between humanist Marx and Gorz where does Gorz's departure from Marx and his alliance with Bookchin lie? The critical difference is that labour in humanist Marx is conceived as being a social activity but Gorz's analysis gives every indication, as we shall try to demonstrate in the interpretation that follows, that it is restricted to a purely technical activity between a producer and nature and this technical activity is necessarily conceived on the model of individual producer and nature or, in even stronger terms, an asocial individual producer and nature. Gorz's free subjectivity is realised in this technical form of labour. The human being is truly free when he posits objectives and achieves them without any external constraints or control in the process of production.

The distinction between the sense of labour of humanist Marx and humanist Marxism on the one hand, and that of Gorz on the other, is not a judgement, at least for the moment, on their relative validity in regard to some valid standard or theory of history. The more relevant question is rather how can a technical conception of labour find its way into a theory which purports to be a critique of relations of exploitation and domination and of human liberation? The question, in other words, is how labour, as technical or 'instrumental' activity, becomes the basis of a theory of human essence?

GORZ'S PERIODISATION OF CAPITALISM

If one were to read Gorz as being a Marxist and to interpret basic notions and formulations including 'labour' as being Marxist notions, one would invariably point to theoretical incoherences and contradictions in his analysis of capitalism, his proposal for socialist revolution and various other themes, particularly as regards the different phases of his work from *Stratégie ouvrière et néo-capitalisme* to *Farewell to the Working Class*. Our hypothesis reveals, however, that it is not contradictions and incoherences which characterise his work but rather an internal theoretical coherence and continuity. Our hypothesis is that Gorz limits his notion of labour to a technical activity between producer and nature in which posited objectives are ones of self-sufficiency and the means of achieving them are free of social mediation. The coherence which this hypothesis reveals is one which derives from an underlying premise that solitary self-sufficiency inheres in human nature. This premise turns Gorz's critique of capitalist social relations into a critique of social relations as such. Let us examine the sense in which our hypothesis captures the underlying structure of Gorz's analysis of capitalism.

Gorz proposes that the history of capitalism comprises two phases: the earlier manufacture phase and the later phase of automation. In this very distinction, Gorz accords determining properties to productive forces rather than to productive relations. This is consistent with his notion of labour as a technical activity. But it is in his specific analysis of each phase that we can identify more clearly this sense of activity. In order to facilitate our seeing this we must follow Gorz's depiction of capitalism beginning with the manufacture phase.

In the manufacture phase, the worker, according to Gorz, is 'estranged' from his work in the actual work process and from the object he produces and this very denial to the worker of any determining or commanding role in the work project is congruous with a rising rate of productivity. In this phase, capitalist domination is objectively compatible with capitalist productivity for it favours a form of organisation of labour and labour process in which the capitalist relations of production are reaffirmed. The labour-power of the worker, upon exchange for a salary, becomes use-value for the capitalist and as such his property to use and command as he sees fit in the pursuit of the goal of accumulation. To this end, the non-producer seeks the most efficient way for maximum productivity involving a particular combination of labour of other workers in the organisation of the labour process. The labour-power of one labourer is a mere quantity without any intrinsic quality and uniqueness in itself. It is but a quantum of physiological energy severed from the worker, as the latter does not posit the objectives of labour. It is combined with other quanta of labour in a multitude of ways which the capitalist himself deems appropriate to maximise his own objective of rising capital accumulation. Labour is alienated and dehumanised as it is severed from the act of positing goals and the act of developing and applying the means to achieve them in the actual work process. In the manufacture phase, Gorz contends that,

> In the great majority of workers, labour power was a pure quantity of physiological energy and this undifferentiated quantity, indiscriminately interchangeable with that of any other worker, did not have any value *in itself*: its worth owed to the fact that

> it could be used and combined *in exteriority* (en extériorité) with other quantities of human energy, that is, to the extent that it was determined by the boss and his representatives, and alienated in a product or a productive process from whose objective he was removed. The worker was supposed 'to work' not 'to think'; there were others in charge of thinking the relationship between his labour and that of others. In brief, the alienation of labour and its dehumanising quality had a natural basis in the division of labour and in the process of production.[2] (author's emphasis)

This depiction of the manufacture phase of capitalist production is rich with theoretical implication. We can detect in it Gorz's anthropology as one which posits human nature as free and autonomous expressing free subjectivity in productive activity. This productive activity, however, is not a social activity as it is in, say Lukács, but rather a technical activity of direct exchange with nature. The first indication of such an anthropology is Gorz's assessment of productive forces as containing a rationality distinct from that of productive relations and, more importantly, as being determinant, in the last instance, of productive relations. Indeed, Gorz presents the manufacture phase of capitalist production as a phase of compatibility between two rationalities – that of the capitalist relations of production and that of the capitalist labour process. This compatibility is determined by the forces of production themselves, in that their level of development is objectively conducive to the capitalist relations of production.

Such a proposition of compatibility allows Gorz to make a number of further claims. First, the estrangement of work in manufacture is not only the form in which capitalist domination is exercised but the form in which domination is compatible with productivity. Second, the relation between productivity and exploitation is reinforced by the minute division of labour and specialisation of function where the organisation of labour remains open to external determination. Third, the division of labour is not only a means of severing the worker from his labour in a total labour process by emphasising his ignorance and incomprehension of the total project but is, at the same time, a means of assuring domination through external combination of labour in a way that is not antithetical to efficient production and hence to the capitalist imperative of continual increase of the rate of exploitation.

It would seem that in Gorz's analysis that which 'negates' man's nature is more the powerlessness in the determination of the work process and less the logic of capitalist relations, i.e. the private appropriation of socially produced surplus-value. This comes out clearer in Gorz's analysis of the automation phase of capitalism. For in capitalist automation Gorz sees the beginning of a process of reappropriation of an 'alienated subjectivity', or the expression of man's essence, and he sees this as constituting labour's regaining a control of the work process. The labour to which Gorz refers is, of course, the technical scientific labour which, in his view, becomes free of technical control by virtue of its technical expertise. For Gorz, in other words, specialised labour, alone, understands the complex intricacies of highly advanced automated technology and thereby acquires monopoly of control of the work process precluding capitalist intervention in the operations of work.

In this account, Gorz equates technical understanding of the work process with free subjectivity, a relation which his anthropology allows him to draw. If the productive activity which expresses human essence is one of direct relation with nature, this productive activity is a purely technical one and the freedom of the direct producer consists of a sufficient expertise of its necessary technical operation. Of course, Gorz cannot claim a 'free subjectivity' for specialised labour in the automated phase of capitalism without claiming that it has a monopoly of technical expertise over the capitalist. For without such a monopoly claim, the capitalist could intervene in technical operations and determine the work process as he does in manufacture.

We can now begin to see the additional element of Gorz's anthropological theory of human nature, namely asocial self-sufficiency. Gorz is able to read a promise of liberation into specialised labour in capitalised automation and thereby to ignore the capitalist relation of private property because man's nature, in his conception, is solitary producer. That which negates human essence here is the social mediation of the productive activity rather than the private form of surplus product, i.e. surplus-value. How else could Gorz contend that specialised labour becomes sovereign in capitalist automation? Let us follow his construction of capitalist automation in order to see clearer these theoretical implications embedded in it.

Gorz's reconstruction of automation in capitalism depicts a changed relation between productivity and exploitation which is, for Gorz, a change in the relation of struggle between classes and suggests the form of radical action which a socialist project should pursue in transforming capitalist social relations. He claims thus that in the phase of automation, productivity can no longer be assured by a division of mechanical skill controlled and coordinated by the capitalist. For automation[3] requires highly synthetic knowledge of a whole range of facets of a labour process of which the capitalist is ignorant and which he, in his lack of understanding, cannot control in order to increase productivity. Labour, alone, as highly specialised and comprehensive, can judge the requirements for an increase of productivity. On the level of productive forces, thus, automation involves a specialised labour with comprehensive, synthetic knowledge; on the level of productive relations it involves a freeing of labour from capitalist control in the labour process. It is important to emphasise this last point for it is, in fact, the labour process on which Gorz concentrates his analysis and in doing so he tends to abstract the capitalist labour process, in a more restricted sense of technical labour, from the capitalist productive relations in which the socially produced surplus-value is privately appropriated. It is because Gorz makes this abstraction that he can claim that in the automated phase of capitalist production, labour begins to discard the alienated form it assumes in manufacture and acquires a form more akin to its essential nature, namely the retention of its essential subjective moment in the determination of the labour process. Here, labour determines itself as opposed to being externally determined or controlled. 'Le travail lui-même ne peut plus être évalué économiquement, par le fait qu'il n'est plus une quantité de temps et d'énergie, une marchandise indifférenciée (Labour itself cannot be evaluated economically in virtue of the fact that it is no longer a quantity of time and of energy, an undifferentiated commodity.)'[4] Labour-power, claims Gorz, is not separated from the worker but reunited with him as he assumes control of the mode of its application. Gorz derives this unity from his conception of labour as technical activity which limits him to the capitalist labour process abstracted from capitalist relations. Let us follow Gorz further in his depiction of automation.

Gorz attributes this 'freeing of labour' not to a good will of the capitalist himself to allow the worker to assume charge, but rather to the nature of the productive forces themselves. According to him, labour is in the best position objectively, from a technical point of view, to judge the manner in which to act in relation to the imperatives of the project. Furthermore, this technical judgement, in reverting to the worker, alters the character of the relation between capitalist and worker. The capitalist no longer combines a number of distinct quanta of externalised labour-power determining its manner of action. Rather, labour remains the quality of the labourer as he himself enters work projects in collaboration with similarly discriminating labourers.

> It is still *labour-power* which tends to escape quantitative evaluations for, qualified workers are no longer, and will be less and less, carriers of interchangeable physical energy, the force of which has value to the extent that it is used and alienated by the buyer who combines it with other undifferentiated forces in exteriority (en extériorité).[5] (author's emphasis).

We see now that that which the worker is 'liberated' from, in Gorz's construction, is external supervision of a technical nature. The worker himself becomes supervisor owing to his specialized, synthetic knowledge and his objective position in the work process in collaboration with others. He assumes responsibility for the operation of a task. He is master of this technical activity and indeed, it is in virtue of this that he becomes 'subject'. Gorz summarizes the new form of labour generated in automation thus:

> It is really impossible to command a qualified worker in advanced industries – he is at the same time labour-power and he who commands it. In brief, he is a *praxis-subject* cooperating with other praxes in a common task which rather imperative orders coming from the top could only disorganise. Here, the worker is an integral part of his labour-power. It is no longer possible to quantify the latter dissociated from the former, both have the same human autonomy.[6] (author's emphasis).

Gorz's abstraction of the productive process from the productive relations is not, of course, a literal one. That is, he does not abstract the latter from the former as to ignore productive relations. The abstraction rather is a theoretical one in which Gorz makes the productive process determinant of productive relations. It is thus that he can claim that the new form of productive process strikes a fatal blow to the hierarchical structure of industrial forces of production, namely, to the externalisation of labour-power and its control by the capitalist in the pursuit of his private goals of capital accumulation. Indeed, he argues that hierarchy, in the form of externalised labour or alienation, has a natural basis in the manufacture phase of productive forces but is replaced in the automation phase by a sovereign 'praxis-subject'. This makes for a new contradiction in the relations of production, according to him. The scientific and technical workers become sovereign in the work process as they, alone, comprehend it and this comes into contradiction with the externally-determined aims of their work, i.e., aims set by the capitalist. This contradiction, argues Gorz further, has already become manifest in certain instances where workers in whole branches of production have expressed the practical impossibility to function within a structure of capitalist management.[7]

The significant point here is that Gorz derives this contradiction as an effect of productive forces. He can make this derivation only if he attributes productive forces determining properties over productive relations. Gorz places, on the one hand, sovereign, creative labour objectively determined by the automated form of the forces of production and, on the other hand, the objective of capitalist labour, i.e. private appropriation, determined by the capitalist relations of production. Furthermore, he makes this contradiction a derivative of the productive forces by claiming that it is a contradiction between the exigencies of productive forces in the form of labourers with synthetic, holistic, technical and theoretical knowledge, and exigencies of capitalist relations, namely, the continuing increase of the rate of exploitation. On the one hand, the inherent logic of the productive forces tends towards the release of domination as external control of labour and the increasing assumption of control by the scientific-technical labourers, and on the other hand, the increasing domination of work by the capitalists through the greater and greater division of labour and specialisation of function and thus fragmentation of labour.

Gorz abandons this thesis in his later work and along with it the thesis of revolutionary potential of specialised technical labour. He retains, however, the theory on which it is based, namely that productive forces are determinant of productive relations and, more fundamentally, that sovereign activity is technical activity as a direct productive relation between producer and nature. It is this theory which prompts his 'adieu' to the working class and which guides him to a new thesis on liberation, that of self-management. As this thesis is central to his socio-ecological theory and as it is most revealing of the asocial quality of the technical activity underlying Gorz's anthropological theory of human nature, we turn to an examination of it in the following section.

SELF-MANAGEMENT: GORZ'S FORM OF LIBERATION

The thesis that one common rationality underlies both the forces and the relations of production marks all of Gorz's later critical analysis of capitalism including his critique of the proletariat and in particular his theory of self-management. This thesis leaves unmodified the assumption of productive forces being determinant of productive relations. The common rationality thesis simply contends that capitalist productive forces are structures of alienation. What must be changed above all are capitalist productive forces, for advancements on this level can no longer hold a promise of liberation for Gorz. The organisation of labour, the technics of production as well as science, itself, 'portent l'empreinte des rapports de production et de la division du travail capitaliste dans leur orientation, leur découpage, leur spécialisation, leur pratique et jusque dans leur langage.'[8] (convey the imprint of the relations of production of the capitalist division of labour in their orientation, their division, their specialisation, their practice and even in their language.)

Whereas previously the highly specialised technical labourers were singled out as being engaged in a praxis in which they were subject, now they are seen as occupying positions which contribute to the reproduction of capitalist structures. They occupy command positions in the capitalist, hierarchical division of labour, dictating the labour of manual and less specialised labourers, but remain within the structure of alienated labour insofar as their own intellectual labour is determined by capitalist relations of private property.

> If technical-scientific workers and manual labourers are in fact situated in the same way *in relation to capital*, they are not situated in the same way *in relation to each other*: as long as technical-scientific labour and manual labour are driven in a *parallel but separate way*, the technical-scientific workers produce the means of exploitation and oppression of manual labourers and must therefore appear to be agents of capital, whereas the manual labourers do not produce the means of exploitation of technical-scientific workers. The relation between them, *where it is direct*, is not a relation of reciprocity: it is a relation of *hierarchy*.[9] (author's emphasis).

Indeed, 'les travailleurs de la science et de la technique ont, *au sein* de leur fonction technico-scientifique, la fonction de reproduire *les conditions et les formes de la domination du capital sur le travail*'.

(*Inherent* in the technical-scientific role of workers of science and of technique is the function of reproducing *the conditions and the forms of domination of capital on labour.*)[10] (author's emphasis). This is so precisely because of 'le souci du capital de dominer et de contrôler le travail vivant de manière à en tirer le maximum de surtravail. (the concern of capital to dominate and control living labour in a way as to draw the maximum of surplus labour.)[11]

In the light of the common rationality thesis Gorz reinterprets the relation between productivity (efficiency) and exploitation in capitalism. Previously he claimed that productivity is maximised in capitalist relations in the phase of manufacture, but now holds that productivity in capitalist relations of exploitation in general is a form of limited *not* optimal efficiency. He appeals to the Chinese experience for an empirical corroboration of this proposition, contending that here, creative labour in cooperative, collective organization makes optimum use of material resources and thus makes for optimum productivity. He concludes thus that this is greater productivity, higher efficiency than that produced in relations of exploitation.

> Maximum physical productivity is simply finding the conditions which permit labourers to produce the greatest quantity possible of given products by applying optimal energy in the most efficient and rational way possible.[12]

Indeed, capitalist productivity can only mean, according to him, maximum quantity of products with the maximum human energy obtainable for the minimum salary.

> Maximum productivity from the point of view of capital, is an entirely different thing: it is finding conditions which permit producing the largest quantity of given products *with the maximum human energy possible for the minimum salary.*[13] (author's emphasis).

Domination of work in an organisation involving division of labour and external control of labour is not, according to Gorz's analysis, necessary for maximum productivity as such; it is necessary for maximum exploitation. Efficient exploitation therefore, is the capitalist form of productivity.

The common rationality thesis implies that the level of productive forces cannot provide a thrust for change of capitalist social relations but it does not deny productive forces determining properties. It is precisely because it claims such a character for productive forces that Gorz's proposal for revolutionary change is a proposal for change at the level of productive forces. The revolution of self-management is a revolution of technology. The change it calls for is a change from a productive activity which is socially mediated, in this case by the capitalist relation of private property, to a productive activity in which the direct producer reproduces his own self-sufficiency. The direct producer develops tools and skills that he alone can command in order to achieve the objective he himself defines.

This revolution has much in common with Bookchin's revolution of self-activity. Such a revolution implies the negation of social mediation of the productive relation between the direct producer and nature. It is a revolution of independent, autonomous direct producers who asocially reproduce the conditions of their self-sufficiency. This is the essence of Gorz's form of social change and the essence of his socialist, libertarian society. It is a society of solitary, self-sufficient individuals which reverts to a state of nature. The form of harmony which is established is that which derives from a negation of social mediation between individuals.

We find the same logic in Gorz's critique of the proletariat. His critique of the working class as it has evolved in capitalism is based on his common rationality thesis. While it appears on the surface that in this critique he has modified his position regarding the relation of determination between productive forces and productive relations making the latter determinant of the former, a closer look reveals that he continues to attribute determining properties to productive forces. Gorz advances his critique of the working class as a critique of capitalist relations for having produced not a revolutionary working class but a group of ignorant workers, exerting quanta of energy mechanically and requiring that the articulation of individual worker energy be external.

> The *collective* power of a class able to produce the world and its history has not been transformed into a subject conscious of itself in each of its individual members. The class that collectively is responsible for developing and operating the totality of the productive forces is unable to appropriate or subordinate this totality to its own ends by recognising it as the totality of its own means. In work the collective worker remains external to the living workers. Capitalist development has endowed the collective

worker with a structure that makes it impossible for real, flesh-and-blood workers either to recognize themselves in it, to identify with it or to internalise it as their own reality and potential power.[14] (author's emphasis).

To his mind, the present form of the proletariat mirrors domination and is as such improperly constituted to lead a revolution and establish a liberated, domination-free society, for 'workers' capacity to recognise the difference between their objective position as cogs in the productive machine and their latent potential as an association of sovereign producers is not inherent in the proletarian condition.'[15] What seems to be more likely, for Gorz, is that the proletariat loses its class characteristics and becomes a non-class in capitalist relations. This claim can only be made if sovereignty or freedom is identified as being synonymous with control of technical activity and this latter as expressive of human essence. The human being is free when he or she determines work in this technical sense. The equation of technical work and human essence, which Gorz makes in his analysis of the proletarian class, leads logically to the conclusion that whoever dislikes work negates his or her essence and as the proletarian workers invariably have an abhorrence for their work in as much as they lack control of it in capitalist relations, they cannot express their essence. The proletariat in remaining confined to a structure of alienated labour in which labour is but a quantum of reified activity loses its class characteristics and becomes a non-class in capitalist relations. 'Just as work remains external to the individual, so, too, does class being', Gorz tells us. 'Just as work has become a nondescript task carried out without any personal involvement, which one may quit for another, equally contingent job, so too has class membership come to be lived as contingent and meaningless.'[16]

It is interesting that in this citation, Gorz replaces Marx's 'species being' of the *Manuscript of '44* with 'class being' as if they were synonymous and as if control of technical labour were the key to this 'being'. Hence, it is not to the proletariat as a class that an appeal for social change can be made but rather to the labourer as individual who, in direct relation with nature, controls the work process. It is the process of self-management.

Before we pursue further Gorz's theory of self-management let us examine his reflections on the relation between capitalist production and ecological destruction as they impinge on his theory of self-management. Unlike Bookchin, Gorz develops a theory of ecological imbalance or ecological destruction in relation to the capitalist relations of production in which he argues that the former are a necessary or structural effect of the latter. Gorz integrates the ecological question into the process of capital accumulation proposing that some ecological factors are constituent of the present form of capitalist economic crisis and that the resolution of the ecological question is inextricably tied to the resolution of the contradiction of the capitalist relations of production. This theory of capitalistically–determined ecological destructions does not however falsify our hypothesis of an asocial self-sufficiency. Let us briefly examine this theory.

The present form of capitalist economic crisis according to Gorz is a dual crisis of overaccumulation and reproduction. That is, superimposed on the crisis of overaccumulation is a crisis of reproduction which, while emerging within the laws of capital accumulation, derives from a fundamental property of nature itself, namely its own absolute finiteness defining absolute limits and absolute scarcity as opposed to the relative character of socially constituted scarcity.

The crisis which Gorz calls overaccumulation is the classic crisis of overproduction which recurs cyclically and is inherent in the very process of expansion, each new crisis engendering a greater increase in the organic composition of capital and a deepening of contradictions. What is qualitatively new here is the ecological component of the crisis of reproduction. Whereas the laws of capital expansion previously presupposed that a scarce good is always producible, they are now confronted with nature's finite limits in which a resource is scarce because it is quantitatively limited in nature. Gorz links, for example, the finiteness of air, water, soil fertility in heavily industrialised areas to the falling rate of profit.

> When air, water, and urban land become scarce, it is impossible to produce greater quantities of them no matter what price is assigned to them. They can only be shared or redistributed in a different way. As far as land is concerned, this means building highrises or underground, or paying higher and higher prices for agricultural land on which to build factories, cities, and roads. In the case of air and water, it means that the available supply must be recycled. . . . The need for such recycling has a precise economic significance: it means that from now on it has become necessary to *reproduce* that which was previously abundant and free. Air

> and water, in particular, have become means of production like any others: industries must now assign a portion of the investments to antipollution equipment in order to restore to the air and water some of their original properties. The consequence of this requirement is a further increase in the organic composition of capital. . . . But there is no corresponding increase in the amount of merchandise produced; the air and water recycled or depolluted by the chemical industry cannot be resold. The falling rate of profit is thus aggravated; the productivity of capital encounters *physical limits*.[17] (author's emphasis).

The exhaustion of mineral deposits has a similar effect on production. The development of increasingly new technologies for the more efficient use of natural resources is the response to the increasingly higher investments required to find new deposits the financing of which translates into higher prices for primary goods. This merely aggravates the problem of the increasing organic composition of capital. The combination of the distinct sources of increasing organic composition of capital makes for a condition in which

> capital . . . encounters unavoidable difficulties in financing further investments – it becomes incapable of ensuring its own reproduction. The replacement of industrial capital . . . can no longer be accomplished by the transfer of a surplus levied upon consumption – the reproduction of the system simply costs more than it yields. In other words, *industry consumes more for its own needs*: it delivers fewer products to the final consumer than it used to. Its efficiency has diminished; its physical costs have increased.[18] (author's emphasis).

Ecological damage deriving from capital expansion must properly become an integral issue of the struggle against capital domination. It finds a solution in Gorz's principle of free labour in which the individual labourer defines objectives as use-value and directly applies tools to produce them. Production for use-value carries with it a double criterion of utility: useful as a function of labour subjectivity and useful as a function of ecological limits.

> Socially produced goods must be available to everyone; their production must not entail the destruction of naturally abundant resources; they must be designed in such a way that, by becoming available to all, they do not cause pollution or bottlenecks which destroy their use value.[19]

It is here, in his proposal for a resolution of the capitalistically–determined ecological imbalance that we find the assumption of asocial self-sufficiency. For the movement out of capitalist relations is a movement of individual direct producers producing the conditions of their individual self-sufficiency.

If self-activity is the link between Bookchin's anthropological theory of the asocially self-sufficient individual, and his political theory of libertarian activity, it is self-management which occupies this place in Gorz's schema. It, too, is an activity of the solitary individual in immediate relation with nature producing his self-sufficiency. As we suggested earlier, this sense of self-management follows from Gorz's implicit anthropological assumption of man as asocially self-sufficient. It is corroborated by the particular construction of 'convivial technology' which Gorz, in fact, borrows from Ivan Illich and integrates into his theory of self-management. What is convivial technology and how in Gorz's schema is it revealing of asocial self-sufficiency?

Convivial technology is first of all an activity of a technical type. It is also an activity in which the individual producer commands tools appropriate to achieve the objective he himself defines. It is an activity in which the individual producer or labourer is subject. Convivial technology thus makes for labour sovereignty as a form of self-sufficiency in the whole labour process from defining objectives to developing and applying tools for achieving them. The group labour which Gorz allows in his theory of self-management can only qualify as convivial technology if the form of organisation of labour is not hierarchical but rather one in which each labourer determines the objectives of the labour project and the technics of achieving them. This means that apart from expressing freely the objectives of the group project, the individual direct producer, or labourer, understands the various facets of the labour of others and is able, in principle, to carry out the total labour process. The individual, as one of a cooperative group, understands and controls individually the tools and the work process. This is necessarily decentralised production. Individual, direct producers or small groups of autonomous, direct producers producing use-value are, then, the essence of emancipatory, free social practice. This is self-management.

THE STRUCTURE OF DUAL SOCIETY AND THE DOMINANCE OF CONVIVIAL TECHNOLOGY

Gorz's theory of emancipation does not end on a utopic note of worker sovereignty as convivial technology. Gorz does not propose simply a world of small, numerous, locally autonomous communities of direct producers engaged in use-value production or free labour to which he refers, following Illich, as autonomous mode of production. He proposes, also following Illich, what he regards as necessary production, exchange-value, heteronomous mode of production – the latter being determined by and subordinate to the former. This involves a dramatic change in the present relation between heteronomous mode of production and autonomous, self-sufficient, local production. The former, which is the sphere of necessity, is subordinate to the latter, which is the sphere of freedom, politics mediating between the two. The state as the sphere of necessity serves the sphere of autonomy in producing and distributing that which each local community cannot produce for and by itself. This heteronomous sphere 'subordinate to the objectives of the sphere of autonomy ... assures the programmed and planned production of everything necessary to individual and social life, with the maximum efficiency and the least expenditure of effort and resources'.[20] It is the sphere of wage labour, reduced in time simply to that which is socially necessary. In contrast, the autonomous sphere is dominant as is the sphere of activities 'which carry their end in themselves'. It is, in Gorz's thinking, the sphere in which the individual labourer can assert his freedom in his self-sufficient labour. From this follows that even the smallest scale of heteronomous production is a necessary evil and derives, according to Gorz, from an irreversible process in the very development of the productive forces: namely (a) the decline of old individual trades which can no longer be learned by the individual labourer for they were never entirely codified or transmitted, and (b) the emergence of narrowly specialised and standardised social skills which are learned by means of formal instruction. The specificity and uniqueness of the form of trade by craftsmen is replaced by a set of social skills which is generalised across great numbers.

The suppression of worker sovereignty persists in this sphere, for social skills are 'predetermined and limited both in their scope and their nature. Instead of belonging to individual members of a "trade", they are the means by which people accede to membership of an economic and social system whose technological development and division of labour remain outside their control. In other words, a "trade" no longer has any use-value for the individuals practising it.'[21] In this structure there can never be effective self-management of big factories, industrial combines or bureaucracies.

> Nothing can support the belief that convivial tools able to assure the autonomous production of use-values can or should be supplied by the autonomous sphere of production itself. Indeed, the more such tools embody concentrated masses of complex socialized knowledge in a form to be easily handled by everyone, the more extensive will be the sphere of autonomy. It is impossible to imagine that telephones, video machines, microprocessors, bicycles or photoelectric cells – all potentially convivial tools that can be put to autonomous purposes – could be produced at the level of a family, a group or a local community.[22]

It is an evil which can nevertheless be put to the service of free labour in the sense of enlarging the sphere of autonomy and thus extending free labour and emancipation. The greater the sphere of autonomous production the freer the society.

This theory of dual society offers a glimpse into Gorz's conception of society of self-management. It is the replacement of 'alienated subjectivity' by 'free subjectivity'. But what can be understood by Gorz's theory of change from alienated to free subjectivity, as this is, indeed, the essential aspect of his theory of self-management? That which seems the most plausible interpretation of this form of change is that it is technical activity. It is a change from one form of technical activity which is either socially or necessarily fragmented, to a unitary form of technical activity which one individual can conduct. The change is from heavy industrial technology to convivial technology. The convivial technology which permits labour independence is a technology in which labour posits a purpose achievable directly by the application of appropriate tools. The link between posited purpose and realisation of purpose is a technical purposive-rational mediation for that which is posited is self-sufficiency achievable only by a technical act of production.

Gorz proposes self-management as a radical political means of changing a society of domination and exploitation to a society of liberation. He proposes this as a means of mobilising resistance to domination and at the same time of achieving individual sovereignty which our analysis shows to be a form of sovereignty in which individuals tend to their self-defined needs without assistance or hindrance from others. But Gorz's political theory does call for preserving some non-convivial technology. Some advanced technology which cannot be accommodated to the model of solitary individual self-sufficiency finds a place in the new society. We see Gorz taking advanced technology and convivial technology to distinguish between two spheres or, what he calls, modes of production – heavy technology being the heteronomous mode of production and convivial technology being the autonomous mode of production. He justifies the heteronomous mode, however, as only a means of providing and sustaining conditions for the reproduction of convivial technology. It should always be in the service of or subject to the exigencies of convivial technology.

Gorz proposes a particular relation between the two technologies. It is a relation in which the needs of autonomy prevail and heteronomy merely serves the objectives of autonomy. This relation holds the key to Gorz's notion of state in such a society – a notion which closely approximates that of Bookchin. Gorz defines politics in this society as the domain of articulation between the autonomous and heteronomous production. It is a site of debate and free expression of both the demands of autonomy and 'the imperatives of technicity'[23] with the goal of defining objectives of government and means of achieving them.

What form can this debate take? What can we understand by the substance of this debate? The answer is to be found in Gorz's notion of free subjectivity as technical labour. It is a debate around the singular issue of maintaining conditions in which such worker sovereignty is dominant. The political debate between individuals can only mean a debate between self-sufficient individuals about the maintenance of conditions of this self-sufficiency. The debating individuals can only be asocial individuals with a common interest of self-sufficiency, an interest free of internal contradiction.

Gorz's state thus is a state of unity of interest of individuals – a harmony which, as in Bookchin, derives from the severance of social structure. For as we have seen all along in our analysis of various aspects of Gorz's theory, his notion of labour is a notion of technical activity limited to purposive–rational action and the agent of this instrumental action can only be the asocial individual. The subjective moment as an act of consciousness in positing objectives is the asocial individual's act of consciousness in setting a purpose of reproducing his self-sufficiency in immediate relation with nature.

The individual is subject in essentially a technical operation of labour. Gorz's cooperative group is nothing but an aggregate of such individuals, each of which controls the whole series of technical manoeuvres of more extensive work projects. This makes cooperative labour simply an extension of individual labour. Gorz's solution to social hierarchy and division of labour thus is situated entirely on the technical level of labour. Indeed, social division of labour is reduced to technical division of labour in which individual self-sufficiency as individual mastery of tools is the solution to social relations of domination and exploitation.

Gorz's theory of self-management and his dual society come remarkably close to that of Bookchin's eco-anarchist utopia. Indeed, in many important respects there is convergence. The terminology may differ but the proposed form of liberated, domination-free social relations is the same. Bookchin's free individual is essentially the same as Gorz's sovereign labourer. Bookchin's decentralised, locally autonomous, self-sufficient community is Gorz's autonomous mode of production. For both Bookchin and Gorz, this form of communal productive relation replaces hierarchy and centralisation as forms of domination. Bookchin's radical action as the acquisition of skills of self-sufficiency and the assertion of individual autonomy is Gorz's radical action as assertion of worker subjectivity in use-value production. For both, this action begins in civil society and constitutes the form of revolutionary change toward emancipation.

The convergence of the two theories owes, of course, to the shared notion of the individual subject. We examined in the previous section the asocial individualist basis of Bookchin's theory. It is the asocial individual in consociation with nature who is the subject of action. The locus of subjectivity in Gorz is also the asocial individual. Neither retain a liberal notion of the individual as private individual. The asocial individual, here, is the depository of subjectivity. He sets objectives and achieves them in direct immediate activity with nature. The individual, as master of his own fate, of his own action, of his labour, is a form of freedom as negation of any social mediation. It is not simply a negation of an exploitative form of social mediation where one determines and appropriates the labour of another.

Gorz's analysis is an individualist adaptation of Marx's theory of labour including its alienated form in capitalist relations. The direct producer is conceived in essentially singular, solitary terms, not in social terms. The implicit model of Gorz's production for use-values and his convivial technology is the solitary individual producer. The use-value he produces is use-value for him. It is not merely an objectified form of his own free subjectivity; it is also the objectified form of his subjectivity as dictated by natural laws. He produces those use-values that sustain his autonomy as self-sufficient individual producer.

The convivial technology which Gorz proposes is limited in scale to that which a single direct producer can control. Even where a tool is of such magnitude requiring a group of producers to operate, Gorz's individual producer must have the capacity to understand all the various facets of the total operation of the labour process. It is this synthetic, holistic understanding of the labour process that characterises the interrelation between labourers of a team as cooperative and not hierarchical.

An implicit model of individual action underlies, as well, Gorz's conception of heteronomous mode of production. The individual, self-sufficient producer is the point of reference for the heteronomous mode of production. This is the sphere of necessity as opposed to the sphere of freedom, that being the autonomous mode of production. But by virtue of what is one necessity and the other freedom? That which makes one the realm of freedom is precisely the capacity of the solitary individual direct producer to posit his goals and to apply appropriate tools to achieve them. The solitary, individual producer is in control of the technics of production. The sphere of necessity is characterised by such means of production as to make it virtually impossible for the solitary individual labourer to know and command the various facets of the technics of production. Here, the individual labourer is obliged to subscribe to directives issued elsewhere and this makes for a hierarchical relation or a relation of domination, in Gorz's terms. Of course, it is not the same relation of domination as that in capitalist relations of production where the objective of production is private appropriation.

The essential difference between necessity and freedom thus is clear. In the former the individual labourer as a solitary, singular producer does not command the labour process from positing to achieving values owing to the magnitude of scale of the means and technics of production. In the latter the individual labourer as solitary, singular producer is in control of the entire work process necessary for achieving predefined objectives.

In the light of these considerations, the term eco-Marxism is a misnomer in depicting Gorz's theory. Although Gorz's general theory is inspired by certain Marxist propositions, it departs from Marxism in the most profound sense. Whereas Marxism is a theory of social determination with an underlying social ontology, the theory of Gorz is a theory of physical determination. Organic nature is the basis of determination of the solitary individual. The subjectivity of solitary self-sufficiency is nothing other than a reason determined by the organic laws of nature. Natural laws determine the subjectivity of solitary self-sufficiency and herein lies the affinity of Gorz's theory with the theory of eco-anarchism.

The implications and conclusions of Bookchin's and Gorz's theories regarding ecology are evident. The individual is united with nature. The social evaporates away. Nature claims the human individual and the organic laws of causality with their own means of regulating ecological equilibria prevail.

It is doubtful, of course, that the similarity between Gorz and Bookchin is fortuitous as if a pure coincidence. A hypothesis of accidental convergence cannot explain away their common anthropology which indeed does determine much of the course of each theory. This anthropology lends a coherence to each theory as a whole as well as to its various elements which on the surface appear incoherent and contradictory. This is not to say that a common anthropology does not allow considerable variations as indeed, we find that these two theories differ in many respects. More important than to identify in their basic assumptions or anthropology the point of convergence is to explain this common anthropology itself. Here hypotheses of chance, accident or pure contingency prove to be still weaker. They lead only to an impasse denying any form of systematic determination.

The anthropology of private or 'possessive individualism' underlies liberal theories from Thomas Hobbes to John Stuart Mill and we have come to understand its full import not as a fortuitous or purely contingently determined assumption of the nature of the human being which the liberals somehow share. We have come to understand it as a bourgeois anthropology determined by the capitalist relations of production. Asocial anthropology is similarly determined as we discussed earlier, but in its determination it is a petit bourgeois assumption which marks the limits and paradoxes of petit bourgeois radical action.

6 Leiss's Critical Theory of Human Needs

The contribution of William Leiss to socio-ecological theory has much in common with that of Bookchin and Gorz. They all purport to reveal the contradictions of social relations of exploitation and domination and to point the direction to a just society and sound ecology. Leiss focuses his critique of exploitative and repressive social relations of the capitalist form in particular, on the level of exchange of commodities, i.e. the level of circulation of capital. He does this in the belief that the crisis of society and ecology is generated by expanding consumerism which he sees as an aberration of human needs and as the source of ecological destruction. His critique of consumerism and indeed his theory of conserver society to which the critique takes him assume an anthropology of human nature remarkably similar to that of Bookchin and Gorz. We shall try to demonstrate in this section that it is another variation of the anthropology of asocial self-sufficiency. We shall argue the sense in which this conception of human nature underlies his critical proposition of consumerist human needs and his programme for a conserver society.

STRUCTURE OF HUMAN NEEDS

In *The Limits to Satisfaction* Leiss tackles the difficult question of the relation between human needs and their satisfaction in various historical forms of social relations casting a most interesting and original light on this question. His objective is two-fold: (a) to determine the general structure of human needs in order to capture the specific relation between needs and their satisfaction in historical forms of social organisation of which he identifies four distinct types, and (b) to arrive at a general social theory which, while revealing destructive consequences of social contradictions on both society and organic nature, may suggest avenues for change to restore ecological order and create conditions for human development.

What Leiss proposes in *The Limits to Satisfaction* is a general, comprehensive social theory even though, by Leiss's own admission, it focuses on some aspects of human behaviour and social relations to the relative neglect of others. The focus is consumption as the whole social process which culminates in the human act of individual consumption.

What explains the centrality of consumption and the 'peripherality' of other social relations in Leiss's theory? This question really relates to the theoretical ordering of issues, questions, themes, relations relevant to a general social theory. The particular ordering in Leiss's theory which accords centrality to consumption owes not to the requirements of general social theory as such but rather to the particular theoretical problematic that regards consumption as the key to human nature as an ontological component of being.

In Leiss's problematic, questions of structure of social relations are posed in terms of relations of exchange. It is an anthropological problematic of 'human nature' where human nature is posed in terms of human needs and their satisfaction. This is a very different matter from an arbitrary highlighting of some aspects to the arbitrary neglect of others. In this particular problematic, relations of exchange are of theoretical importance because they are claimed to hold the key to questions of how human beings organise to assure their reproduction, i.e. questions of social reproduction. In the terms of the problematic, in other words, human needs and their satisfaction and relations of exchange have determining properties and an understanding of their structure is also an understanding of other aspects of social structure. We suggest that this problematic of exchange contains the key to the affinity between Bookchin, Gorz and Leiss, but before we examine the sense in which this claim can be sustained, let us follow Leiss in his construction of human needs and their satisfaction.

The most basic of Leiss's contentions is that all human needs are socially mediated. It is a contention which in the terms posed is not free of considerable theoretical ambiguity – an ambiguity deriving from the problematic itself. What is the form of the social mediation of human needs? When and how does the social intervene in the constitution of needs? In other words, how are needs socially constituted? Leiss provides different and contradictory answers to the question. One is that the sociality of needs is anthropologically determined as social. 'We are innately social animals with an extraordinary capacity for developing those symbolic representations that are summed up in the term *culture*.'[1] Speaking of the complexity of 'human needing' which we shall examine later, he claims that it 'is inherent in the species' evolutionary origins; it is not itself the result of historical or cultural development, but is rather the necessary presupposition for that process. ... The potential elaboration and variation of this innate symbolizing capacity was initially dependent upon the evolutionary success of *homo sapiens* vis-à-vis other *hominid* species and other mammals'.[2]

The sentence immediately following the first quotation distinguishes between spontaneous impulses striving for self-preservation and the representation of these impulses. This implies that the social constitution of needs is symbolic or, we may say, it is a process of ideological transformation of impulses into social needs.

> In the intensive social interactions that have been always present in human societies, men and women collectively interpret, reflect upon, and integrate those otherwise spontaneous impulses that we normally call the striving for self-preservation in a species. Impulses are mediated or transformed by cultural forms into needs, that is, conscious expressions of desires that become 'congealed' in socialization patterns transmitted from one generation to another. The cultural, symbolic, or 'reflective' transformation of impulses breaks the direct interaction between instinctual drives and the means for gratifying them that normally occurs for other species through adaptation to environmental fluctuations.[3]

The social as represented here is no longer innate but rather an intervening element in a relation between impulses and their satisfaction. '[In] the development of tools . . . the direct interaction between impulses and sources of satisfaction is broken; impulses are controlled and consciously directed toward an enlarged field of satisfaction.'[4] In the latter part of the passage, the sociality of needs is no longer ontological but historical and it is this second reading that Leiss seems to prefer.

> One of the outstanding characteristics of human development is this movement away from reliance upon the spontaneous providence of nature for the satisfaction of needs. In other words, there is an increasing intervention in and management of environmental processes inaugurating a reciprocal interplay between symbolically mediated impulses (needs) and organized techniques for provisioning them.[5]

With the transformation of impulses into needs the material component is not eliminated, becoming something purely conceptual or symbolic, but rather a need becomes both material and symbolic. This becomes one basic principle of Leiss's theory of human needs, namely, that human needs have a structure and that the structure consists of material and symbolic aspects neither of which can be collapsed or reduced to the other nor separated from the other. These material and symbolic aspects of needs are referred to as correlates which serve as well to define Leiss's notion of 'state of needing' such that this latter becomes synonymous with needs.

> The term *correlate* seems to be the most appropriate concept in this context for the following reason: although the two aspects of needing can be distinguished for purposes of discussion, they never function separately in the actual process of needing; in other words, the experience of needing is inherently a multidimensional activity.[6]

The claim of indissoluble unity and irreducibility of material and symbolic correlates of needing does not, of course, settle the issue of the relation between them. If the symbolic is not reducible to the material and if the material is not reducible to the symbolic, they are nevertheless conceived as correlates and as such comprise a structure, namely the structure of needing. Correlates that are irreducible but which nevertheless inhere in a structure submit to some form of articulation. They relate to one another though not as pure exclusion nor as pure determination for Leiss tells us that neither is purely autonomous from the other nor is one reducible to the other.

> Neither the material nor the symbolic aspects can be reduced or collapsed into the other; nor in my opinion (contrary to arguments by writers such as Baudrillard) can the two be separated, so that one could claim to detect a 'symbolic order' that is divorced from the sphere of political economy. I believe that the human system of needs in every culture is an indissoluble unity of material and symbolic correlates.[7]

What is the form of the relation between them? How are the two correlates articulated in a structure? There does not seem to be a clear answer in Leiss's theory of needing. Leiss claims only that the relation is complex. But this qualification of complexity of articulation of needs does not settle the issue, for all it claims is that human beings as a species are predisposed to provide symbolic mediations of needs.

> There is no movement from simplicity to complexity, only a progressive unfolding and enrichment of the symbolic mediations subsisting in reciprocal relation with the continuous reinterpretation of needs. The potential elaboration and variation of this innate symbolizing capacity was initially dependent upon the evolutionary success of *homo sapiens* vis-à-vis other *hominid* species and other mammals.[8]

It is in terms of this structure of human needing, however obscure and ambiguous a structure, in that we know only that it comprises material and symbolic correlates and that the articulation between them is complex, that Leiss interprets the various forms of social formations. He proposes a four-fold social typology which he refers to as forms of human ecology. These are the human ecology of hunting and gathering; the 'small-scale permanent settlements, initiating the production of the means of subsistence; large-scale settlements and then civilization, with considerable spheres of exchange and the division of labour; capitalism and industrialization, in which the high-intensity market setting emerges after some time'.[9]

The notion of human ecology or ecological pattern in reference to social formation provides some clues to Leiss's notion of structure of material and symbolic aspects of human needing. It seems that the material aspect or correlate of human needing relates to a 'material exchange' with organic nature in which nature (or objects of nature) are worked upon to form objects of use and that it is this production for use which becomes symbolised in some form or other. How else can we understand a statement such as the following?

> The refinement of desire and the environmental manipulations that are associated with the sedentary state opened new vistas for the instrumental applications of the symbolizing capacity in building, crafts, agriculture, and so forth.[10]

In this passage Leiss discusses the material exchange with nature as a relation of production for use. This material productive relation for use, according to Leiss, acquires an additional form, a symbolic form in such a way that the material becomes inseparable from the symbolic. What is social here is the symbolic interpretation of the material productive relation for use. If this is so, this passage implies that the material productive exchange with nature for use is not a social relation at all. It is social merely in its symbolic aspect, i.e. as an interpretation of the material. This reading of the passage seems to be borne out by another passage discussing a distinction between production for use and production for exchange, part of which we have already cited above.

> One of the outstanding characteristics of human development is this movement away from reliance upon the spontaneous providence of nature for the satisfaction of needs. In other words, there is an increasing intervention in and management of environmental processes, inaugurating reciprocal interplay between symbolically mediated impulses (needs) and organized techniques for provisioning them. In the highly developed stages of this process the division of labour emerges; the individuals engaged in specialized production cannot satisfy all their needs directly but must depend upon regular exchanges of goods in the market place.[11]

Here, the productive relation between producer and nature becomes social only when 'individuals engaged in specialized production cannot satisfy all their needs *directly* (my emphasis) but must depend upon regular exchanges of goods in the marketplace'. The (social) relation is social as a relation of exchange.

Leiss's discussion of satisfaction of needs in which context he advances his reading of Marx, lends further supporting evidence to our interpretation. Leiss conceives of production essentially as a means of satisfaction of needs of which he proposes a four-fold 'pure type' classification: 'production solely for use; production primarily for use, but including limited exchanges, in small-scale social units; production primarily for use, but including limited exchanges, in larger social units; and production primarily for exchange'.[12] The first type occurs in hunting-and-gathering societies and here the relation of production appears to be not a social one but rather a relation between producers/users and nature in which they individually obtain need satisfaction.

> The members of the social unit normally produce all the artifacts that are necessary for the provisioning of their needs. This situation changes only when such peoples come into contact with others that are more technologically advanced. When this occurs they may begin to trade for items which they cannot fashion for themselves.[13]

The relation to the means of production, 'the artifacts', is not depicted as a social relation. If it is assumed to be a social relation, the form of this social relation is left unnoticed *as if it were not problematical*, a factor determined by the very problematic of exchange as distinct from a problematic of production.

The second type of production for the satisfaction of needs in Leiss's typology is production primarily (but not exclusively) for use in small-scale social units which occurs in small-scale settled societies. Here, production for use 'is organized around the extended-family unit which generally provides most of what is required for subsistence',[14] although some exchange occurs 'within and among *cultures* of this type' (my emphasis). The important aspect of this exchange is not only that it is limited but it is governed by rules and norms which classify objects and specify which exchange is sanctioned (i.e. within or between classes of objects.)

The third type occurs in large-scale permanent settlements where, we are told 'the majority of the population continues to depend primarily upon production for use in the satisfaction of their needs'.[15] Exchange is mainly non-market and state-managed, although some limited market exchanges do occur. European feudalism is depicted as being one of these and as having the unique distinction of having 'prepared the seedbed for the emergence of capitalism, industrialism and unlimited market exchange'.[16] What is interesting in this account is the depiction of the transition.

> Here all the 'factors of production' – land, labour, and capital accumulation – were transformed into an open-ended series of market transactions that promoted technological innovation and the division of labour at an accelerating rate. For the first time a market *economy*, as opposed to market exchanges, came into being. Production for use gradually dwindled, being confined largely to the 'backward' sectors of the economy (such as farming), and the majority of the population became dependent upon the purchase of commodities for satisfying their needs.[17]

How is it possible to conceptualise this process as occurring at the level of exchange? How is it possible to conceive of 'land, labour, and capital' as uniquely relations of exchange promoting technological innovation and a division of labour at an accelerated rate, or at least as relations determined in the final instance by relations of exchange? The answer is not difficult to find. It is a construct of production from the perspective of consumption rather than consumption from the perspective of production. The notion of mode of production thus, is replaced by the notion of pattern of need-satisfaction and this is the key to Leiss's critical reading of Marx and to his reinterpretation of the theory of historical materialism in terms of a theory of need-satisfaction. Leiss provides his interpretation of Marx from the perspective of his conception of commodities as need-satisfiers in the context of his fourth pattern of need satisfaction, namely, 'the high-intensity market setting'.[18] Here, commodities are examined as depositories of material and symbolic correlates in relation to the consumer. They are complex objects (as constitutive of particular material and symbolic aspects) emitting messages to the consumer. Leiss's formulation of the structure of commodities and their relation to the consumer depicts this very clearly.

> Commodities themselves here become very complex objects: the ambiguities inherent in the material-symbolic correlates of human needing itself are replicated in these objects. They are not simply material things in 'material-symbolic entities' – that is, things which embody complex sets of messages and characteristics. These incorporated messages offer suggestions to the purchasers of commodities concerning the suitability of the commodities for the needs of individuals.[19]

The 'consumption' perspective overlooks entirely, of course, the relations of production and the sense in which they become determinant of the structure of commodities. What is lacking in Leiss's account of commodities is that particular commodity, i.e. labour-power, which emerges as commodity with exchange-value and use-value just as any other commodity but with the difference that its commoditisation owes to a particular productive relation – the dispossession of labour of the means of production. The productive relation cannot be conceived in terms of need-satisfaction in which the capitalist buys labour-power for use in producing surplus-value, unless, of course, the notion of need-satisfaction becomes equivalent to class interest, which is not what Leiss has in mind. And neither is labour-power equivalent to 'services' as distinct from 'goods' which circulate in the market economy, for this notion too slides over the productive relation and situates the economic relation as merely a relation between object (service or good) and consumer.

Leiss's critique of Marx is inspired and conditioned by his problematic of exchange. He limits his critique to some aspects and specific passages in Marx. In his citation of Marx's account of the change from pre-capitalism to capitalism, for example, Leiss omits in the very citation a critically important part of Marx's analysis, namely, the relation of production. This passage is omitted as if it were not of significance in itself nor to the rest of the passage. Leiss's citation of Marx reads thus: 'The dissolution of all products and activities into exchange values presupposes the dissolution of all fixed personal (historic) relations of dependence of . . . the producers on one another'.[20] Marx's passage reads as follows: 'The dissolution of all products and activities into exchange values presupposes the dissolution of all fixed personal (historic) relations of dependence in production, as well as the all-sided dependence of the producers on one another.'[21] By eliminating the essential factor of 'dependence in production', Leiss loses the point that the fundamental dissolution in question here is the precapitalist relation to the means of production, which is a 'fixed personal (historic) relation of dependence'. This relation of production becomes a relation of 'free' impersonal (historic) relation of dependence in capitalism. In other words, the relation of producer to the means of production in capitalism is a relation of separation of the producer (labourer) from the means of production, access to which can be achieved only by exchanging labour-power for a salary. The exchange is governed by an impersonal factor of maximisation – the seller of labour-power (the labourer) for the highest salary obtainable and the buyer (the capitalist) for the least salary in relation to the use to which he will put the labour-power in order to maximise surplus-value. Here lies the relation of production which renders the capitalist commodity structurally different from those of pre-capitalist productive relations.

This is not a minor point of misinterpretation of Marx. It is rather a revelation of the conceptual perspective for which relations of production are either trivial or determined by relations of consumption. Leiss's reproach of Marx for the sense in which he distinguishes production for 'use' from production for exchange is relevant here. 'A deficiency in Marx's analysis,' Leiss tells us, 'is that in some passages he drew too sharp a distinction between "direct production" for use and production for exchange. Or, to phrase the point more accurately, he emphasised the elements of discontinuity between the two modes of production without drawing attention to the equally significant elements of continuity'.[22] For some supporting evidence of this statement he refers to at least one particular passage in the *Grundrisse*. Leiss cites that

> In production for exchange 'all individuality and peculiarity are negated and extinguished. This indeed is a condition very different from that in which the individual or the individual member of a family or clan (later, community) directly and naturally reproduces himself.[23]

but Marx finishes his sentence thus:

> or in which his productive activity and his share in production are bound to a specific form of labour and of product, which determine his relation to others in just that specific way.[24]

The notion of sharp distinction or discontinuity might appear a little loose in a theoretical sense. This 'looseness' or vagueness is, however, of considerable theoretical significance in the context.

The distinction which Marx draws is a structural distinction of productive relations. It is a distinction between modes of production on the basis of the relation to means of production. The capitalist productive relations are relations of separation of direct producers from the means of production whereas in pre-capitalist modes there is a relation of unity of direct producers to means of production. It is this relation of dispossession or structural separation of direct producers from the means of production in capitalist productive relations which 'negates and extinguishes all individuality and peculiarity' since the sole mediation of access to production is the commoditisation of labour-power. Whether this structural difference between pre-capitalist and capitalist productive relations can be characterised as a difference of revolutionary import or of continuity cannot be judged from the perspective of consumption where Leiss situates the issue. But a conception that production for use is not distant from production for exchange and indeed that the two are related by some notion of continuity is itself determined by a conception of consumption (need-satisfaction) as determinant of production.

THEORY OF COMMODITY

Leiss's critique of Marx sets the theoretical stage for an elaboration of his notion of commodities which while not defined purely and simply as an object of consumption is, nevertheless, determined by the consumption perspective as an object in relation to the consumer. What is of interest to Leiss thus is (a) the structure of commodities as an object of consumption (b) the relationship between this structure and what he calls 'high-intensity market setting' and (c) the relationship between the structure of commodities and the structure of needs, or in other words, how the high-intensity market setting determines these structures and relates one to the other.

A commodity is 'a tangible artifact produced and exchanged for the satisfaction of needs'.[25] Leiss's notion of commodity is more than a description of an object of consumption as a use-value or as particular quality or characteristic or set of characteristics deemed by the consumer to be appropriate to satisfy a need or needs. It is, rather, a relation between product and consumer. The proper relation, according to Leiss, is one in which the consumer can exercise a judgement of the product's suitability for him or her on the basis of (a) his or her 'craft knowledge concerning the quality of things'[26] and (b) his or her 'capacity to make an independent assessment of its suitability for (his/her) needs'.[27] This is the conception of the relation of use-value which is retained and which, indeed, determines Leiss's notion of consumer society as well as his critique of the 'high-intensity market setting'.

According to Leiss, the essential characteristic of 'high-intensity market settings' by which he means advanced capitalism and socialism, is a fragmentation of the structure of commodity as a bundle of qualities or characteristics of an object of consumption and a fragmentation of the structure of needs of the needer. What is this process of double fragmentation of which Leiss speaks and how is it determined by the 'high intensity market economy?'

Leiss's notion of fragmentation of commodities provides a most interesting and insightful theory of the dynamics of marketing and advertising. According to him, the fragmentation or disintegration of structure of commodity consists of the abstraction of the use-value or characteristics from the material object itself and of the implantation of these characteristics into a whole range of other objects such that no object retains a unique value but rather loses its unique or 'independent' identity, so to speak. Furthermore, this process of abstraction and reimplantation is continuous and rapid. It is a process largely propelled by modern advertising which imputes characteristics to commodities in an endless production, combination and promotion of characteristics such that '[c]ommodities are not straightforward "objects" but are rather progressively more unstable, temporary collections of objective and imputed characteristics – that is, highly complex material-symbolic entities'.[28]

The fragmentation of needs, on the other hand, refers to a state of confusion on the part of the consumer between his needs and the means by which they may be satisfied. This state of confusion is actually determined by the market economy such that the fragmentation of needs 'stand in reciprocal relation to the disintegration of the characteristics of objects',[29] and a 'complete identification of needs and commodities'[30] is produced.

The rapidity or intensity of commodity circulation seems to be the key to the double fragmentation. A dialectic is produced between the process of disintegration and fragmentation of commodities and needs, on the one hand, and the process of commodity circulation, on the other, such that an intense circulation of commodities generates disintegration of commodities and fragmentation of needs and this disintegration and fragmentation intensifies circulation. More and more commodities are produced, more and more waste is generated, more and more ecological damage is incurred. An ever greater confusion or difficulty of matching the qualities of needs with characteristics of goods and a growing indifference to the quality of needs results. It is this, in Leiss's view, which constitutes the crisis of our civilization.

Leiss sees as imperative a social change which would rupture the direct relation between needs and commodities. His conserver society is a 'social setting' in which the direct link between needs and commodities is severed. This conserver society is one in which there is (a) reduction of commodity circulation, (b) reduction 'in the economic and political power of multinational corporations', (c) reduction of per capita consumption of energy, (d) marked increase of production for use-value in which individuals directly produce the means of satisfying their needs, including health, lodging, food, etc., (e) an end to social interference in the process of homeostasis of non-human nature, where homeostasis is socially recognized as a non-human need and a natural right.

LEISS'S ASOCIAL ANTHROPOLOGY

It is apparent immediately, that Leiss's conserver society has much in common with Bookchin's eco-utopia and Gorz's post-industrial socialism. The basic feature of all three is the activity of individual self-sufficiency in which the individual posits his needs and satisfies them directly by 'direct production of use-value'. It is this which resolves social hierarchy in a transformation to eco-utopia in Bookchin; it is this which resolves worker alienation and achieves a post-industrial socialism in Gorz and it is individual self-sufficiency, as well, which ruptures the direct link between needs and commodities in Leiss as the individual achieves 'a more direct and personal participation in activities related to the satisfaction of needs'.[31]

Leiss's discussion of his alternative setting and conserver society is not as detailed as comparable discussions in Bookchin and Gorz, and various issues of transition are less well worked out. Direct use-value production is proposed as one component of a policy for social change but the modalities for arriving at such a programme and for achieving it remain vague.

It is possible to note some tautology in the vagueness of the proposals for transition. Leiss contends that there are two 'necessary conditions for the emergence of alternatives . . .: the institutional organizations required to bring about a conserver society; the practical understanding of human needs in relation to the needs of other living entities in the biosphere – indeed, in relation to non-human nature as a whole'.[32] But if we are correct in our interpretation that direct production for use-value is individual self-sufficiency and it is the determining principle of Leiss's 'alternative setting' or conserver society, all that is vague acquires clarity. It is this which casts light, for example, on the full significance of the direct link between needs and commodities in 'high intensity market setting' as a result of a loss of a 'craft knowledge concerning the qualities of things'.[33] The individual's capacity to judge the means of satisfying his needs – needs which he alone defines – is a function of the individual's capacity to produce the means of satisfying his needs. The individual is free and independent in the act of positing his needs and in developing the means of realising them. What is more fundamental than Leiss's assertion of social determination of needs is an anthropological assumption of individual self-sufficiency, the same assumption underlying Bookchin and Gorz.

The individual's loss of capacity (as a craft knowledge concerning qualities of things) to judge the means of satisfying his needs is the form of his enslavement in capitalism. He becomes tied to or imprisoned by commodities. The cutting of this tie is the recapture of the capacity to define needs and develop means to satisfy them. We see then, that the craft knowledge is the know-how of a technical activity of production for use-value. It is essentially a direct, socially unmediated activity between producer and organic nature.

In examining Leiss's analysis of modern society, we argued that he adopts a consumption perspective from which to view the social processes of advanced capitalism. We noted that from this perspective not only are relations of production seen as being determined by relations of circulation but as all commodities are seen as having the same structure, a 'blind spot' is cast on the uniqueness of the structure of the commodity, labour-power. If our interpretation of individual self-sufficiency is correct, some additional light is cast on what we have called the consumption perspective. First, in regards to the capitalist mode of production, it is not a perspective of analysis which purely and simply attributes social relations of exchange determining properties over social relations of production. Neither is it a perspective which claims that relations of exchange are determinant of relations of production in any mode of production or social formation. Indeed, the 'consumption perspective' is not a theoretical construction of social processes of consumption as being determinant of other processes.

How then can we understand the selection of consumption by this perspective as the key to revealing the contradictions of capitalism, 'industrial socialism' or of any pre-capitalist relations of exchange? What it depicts as consumption is a relation between individual and use-value. The relation is conceived as being essentially immediate without social mediation although it is set in a social context with the expressed objective of identifying the 'cultural determinants' or, in other words, the social mediations of this relation.

This interpretative argument seems to be paradoxical for how can we say that the relation between consumer and use-value is conceived as being immediate, i.e. unmediated by anything other than the technical activity of satisfaction of a need, and at the same time how can it be examined in relation to social processes, i.e. purchase or exchange, supply and demand, etc? What we are arguing, however, is not paradoxical if the theoretical construction of the relation between consumer and object is understood (or interpreted) as being an abstraction from social structure and a reinsertion into social structure for the purpose of analysis of this social structure. Abstraction and reinsertion are two theoretically separate movements different from a conception of the relation as a social relation in its very *constitution*.

Second, if our interpretation of individual self-sufficiency is correct, what we have called the theoretical perspective of consumption is shown to be a theoretical perspective not of consumption, even in this narrow sense of socially unmediated relation between consumer and object, but of production and this relation is the relation between individual producer and organic nature in which the producer determines his needs and satisfies them directly or immediately. This relation of production is one and the same with the relation of consumption, namely the individual in 'direct' exchange with nature producing his use-values. What we originally identified in Leiss's analysis as a theory of social relations from the perspective of circulation is really a theory from an anthropological perspective of solitary individual self-sufficiency.

We have already discussed in the previous two sections the social determination of this asocial anthropology and the sense in which it is a petit bourgeois response to the contradictions of advanced capitalism. We discussed as well the limits and paradoxes of individualized direct production for use as a radical reaction to the contradictions of capitalist relations as they are experienced by the petite bourgeoisie. A repetition would be redundant here. What is of unlimited interest and what indeed lends additional support to our hypothesis of the petit bourgeois class character of asocial anthropology is that a number of theories so disparate on the surface appearing to converge only thematically are unified by a deep structure which is the reproduction in theory of the ideology of the social relations of production.

Conclusion

The critical spirit of this essay is not one which seeks to establish the claim that all is well with ecology and that any expression of concern about environmental matters is an unfounded, alarmist outcry. It does not deny that grave ecological problems exist nor does it suggest that our attention should focus away from ecology to more critical issues as do some critics of the ecological movement.[1] The objective of this essay rather, is triple: (a) to examine how each theory specifies its object, ecology, (b) to identify the underlying premises of each theory in its role as basic determinant of internal logic of the theory, (c) to assess each theory as a theory of explanation and a theory of political action. Rather than assuming that ecology merits little concern, this triple objective implies that it is of such critical importance that it is incumbent upon us to understand the sense which these various theories make of ecology and the sense of their call to political action. To understand this is not to dismiss ecology as an issue, although it may involve rejecting some of these theories or parts of these theories as a programme for resolution of social contradictions of which the ecological issue is an effect. Before we begin considering this last point which must be the central question of this chapter, let us assess each of the theories against each part of this triple.

NEO-MALTHUSIANISM: THEORETICAL SPECIFICATION OF OBJECT AND THEORETICAL PREMISE

The object of each theory is ecology but this object receives such differing and often opposing, conflicting treatment that it can hardly be considered the same object. Indeed, it is not the same object, for what we have in these theories is not ecology as a real, concrete object that we may observe as we observe a given tree or a given rock. These theories demarcate a theoretical object and it is the constitution of this object and the political theory to which it gives rise that are of principal interest to us. The theoretical object is ecology in theory varying from one theory to the other. The differences are dependent on the theoretical problematic which specifies the questions that can be raised regarding ecology, the relation between these questions and their specific sense. The problematic is the fundamental determinant of the theoretical object.

The neo-Malthusian problematic poses ecology, or rather the state of ecology, as the effect of a relation between a human population mass and physical, organic nature. This relation is posed as being contradictory. It generates crises of ecology in that the population has a property of limitlessness but nature has a property of limitedness. On the one hand, populations tend to expand numerically at an exponential ratio and to indulge in unlimited activity including economic activity in which nature constitutes the raw material. On the other hand, organic nature tends to inhere to a finite dynamic expressed in the law of matter as neither creatable nor destroyable. It is this imbalance, indeed contradiction, which generates the crisis of ecology as an overtaxing of nature's capacities to provide food, natural resources and to absorb or adapt to multifaceted forms of pollution, including radiation. The crisis, then, derives from excessive human demands which contradict nature's requirements for assuring its own reproduction. Human population in this problematic is conceived both in terms of undifferentiated mass and as an aggregate of individuals of insatiable appetites including but not uniquely the appetite for having children.

The conception of both elements of the relation – human population and organic nature – has an evident quantitative component. Population size, rate of demographic expansion, as well as number of appetites and extent of activity of a typical individual of the population mass – all notions of quantity – provide the sense which underlies ecology as a relation between population and nature. It seems to be a relation of consumption and 'ecology in crisis' seems to be a relation of overconsumption or abusive consumption as in some forms of neo-Malthusianism critical of nuclear technology. The key notions in the problematic then, are quantity, consumption, infinity of human appetites, finitude of nature.

The questions which the problematic invites are questions regarding degrees of activity, extent of activity, quantitative rates of expansion, both demographic and industrial infra-structural, degrees and extent of soil fertility, quantities of natural resources, extent and rates of pollution wastes, demand and supply of food as a relation between numbers of people and food production, this latter being, itself, a function of soil size and fertility, relation between a given form and degree of activity and extent of raw material utilised and so on. What are conspicuous by their very absence from these series of questions and related questions are questions regarding social relations. Indeed, while the neo-Malthusian problematic invites the former it precludes the latter not by some mechanism of awareness and conscious exclusion but rather by its own logic and coherence. It precludes questions of social structure or social relations, of environmental effects of social relations, of the relation between the internal organisational logic of a given form of economic, political and social practice and environmental effects. This problematic situates the contradiction or contradictions of which the ecological crisis is the effect not in social relations but rather in the relation between a socially undifferentiated mass of people, and thus socially unmediated human activity, and nature. The consumption of nature to which the basic relation of this problematic alludes, includes, of course, productive activity but this productive activity is conceived largely as technical activity with an acquisitive objective. That is, the activity is that of the individual seeking to satisfy his infinite appetites by instrumentally appropriate means.

The basic quantitative logic of this problematic invites a particular type of corrective measure to resolve the ecological crisis. If the crisis is basically an imbalance between quantities – larger numbers and greater demand at a greater rate of growth on the side of human population and activity and smaller numbers, lower rate of renewal on the side of organic, planetary materials and capability – the crisis can be resolved by restoring a balance. This can be achieved by reducing demographic numbers and rate of demographic growth, by reducing activity requiring raw materials and by producing wastes to a level comparable with natural capacity and organic exigencies for renewal or reproduction. This becomes the core of the neo-Malthusian political theory.

The neo-Malthusian political theory of restraint thus is dictated by its very problematic. The restraint is either internally imposed by internalisation of values of moderation, frugality, conservation, etc., or externally imposed by political coercion controlling activity to a degree compatible with natural exigencies as determined by scientific knowledge at a given moment of the development of science. This control may involve, as well, the form of technical means to obtain ends such as reduction or elimination of nuclear energy in favour of solar energy. The neo-Malthusian problematic limits the political theory to proposals for restriction. Its very logic excludes change at the level of social relations. Since the logic of social practice is pre-empted from the diagnostic level of analysis of the ecology crisis so is it pre-empted from the prognostic level of analysis for crisis resolution. If, in the logic of the neo-Malthusian problematic, social relations do not figure in the generation of the crisis, they cannot figure in the resolution of the crisis.

While all these notions, questions, relations and normative prescriptions can be understood in terms of the neo-Malthusian problematic and while unraised questions and omitted notions can similarly be understood, how can we understand the problematic, itself? That is, what underlying logic does it imply? What is the premise or basic assumption on which it is based? The postulate which both presupposes and follows from the neo-Malthusian quantitative problematic is that: (a) human beings have infinite, insatiable appetites and are driven to develop and pursue means to satisfy them; (b) to the extent that these means involve organic nature as raw material they are limited; (c) organic nature as a system of interdependent natural processes is a finite system of delicate equilibria easily threatened or disturbed by overuse. The postulate posits a relation between the three elements which is a relation of contradiction. On the one hand, human beings have infinite appetites and indulge in unending and expanding activity to satisfy them, but, on the other hand, the physical organic planetary system which serves as a system of support of this activity is finite and is able to support only a certain amount of activity. The viability of nature is jeopardised by human activity but human activity is dependent on the viability of nature.

This postulate which is part of the neo-Malthusian problematic presupposes something more fundamental. It presupposes a liberal anthropology. It assumes that: (a) a human essence inheres in human beings; (b) that this essence is infinite acquisition; (c) that each individual seeks to realise his essence by developing means which maximise the returns of his activity. This liberal anthropology, clearly of a Hobbesian form, does not lead to an obviously liberal economic, social and political theory of the kind, for example, we find in Hobbes. On the contrary, Hobbesian liberalism remains rather obscure or hidden in neo-Malthusianism. This is so because this liberal anthropological premise is combined with the naturalist premise of organic, physical finitude – a premise for which modern empirical science provides considerable evidence. The contradiction which follows from these two premises together is resolved not in favour of one or the other premise. Neither is the liberal anthropology nor the premise of natural finitude dropped from neo-Malthusianism.

Their retention has a critical effect on all levels of neo-Malthusian theory including the very problematic which specifies the theoretical object, ecology. It has an effect on the political theory where politics becomes a politics of constraint or coercion of the individual to limit his appetites and activity in line with the viability of nature.

If all this is the effect of the fundamental liberal anthropological premise in association with the premise of natural finitude, of what cause is this anthropological premise itself the effect? Is it a scientific, ontological premise or an ideological premise? Before we turn to the question of the structural basis of this liberal anthropological premise let us look, in the way of summary, at the process of constitution of the theoretical object, ecology, in each of the theories examined in the previous chapters and take up the matter of the nature of these premises in a general assessment of the explanatory character of each theory later on.

EXPANSIONISM: THEORETICAL SPECIFICATION OF OBJECT AND THEORETICAL PREMISE

The expansionist theory, as we have pointed out earlier, is not a critical theory of ecology in the same vein as the other theories. It does, however, address environmental and ecological issues and the manner in which the issues are perceived is determined by the expansionist theoretical problematic and the premise underlying the problematic. Since it shares a common premise, i.e. Hobbesian anthropology, with that of neo-Malthusianism it is an important referent for judging the critical import of neo-Malthusian socio-ecological and political theory. It seems, then, that its explicit treatment of ecological matters and its theoretical proximity to neo-Malthusianism in spite of conflicting differences, justify more than a passing reference to expansionism.

Similarly to neo-Malthusianism, the expansionist problematic poses ecology as the effect of a relation between human activity and physical nature. Unlike neo-Malthusianism the relation here is free of contradiction because neither human activity nor nature are posited as inherently finite. Nature is seen as providing the raw material on which human activity expresses man's essence. It grows and evolves in relation to this activity. Since, like neo-Malthusianism, the expansionist problematic takes up the liberal Hobbesian anthropology, the conception of human activity, is one of limitless private acquisition. The Hobbesian human nature of infinite appetites which seeks satisfaction in an activity in which, according to the expansionist problematic, simultaneously unfolds human essence and develops nature in an unending forward movement. Just as in neo-Malthusianism, as well, this problematic poses the activity essentially in Hobbesian terms, namely, technical mastery. Since the subject is unlimited in technical ingenuity and nature unlimited in its receptivity to changes to which it is submitted, ecological imbalances or disorders are not seen as being irreversible or technically uncorrectable. On the contrary, it is in this sense that the expansionist problematic highlights technology as the ecological solution.

This faithful reproduction of Hobbesian assumptions in the expansionist problematic leads to an unmodified Hobbesian political theory, namely a freedom and liberty of the human subject to pursue his privately defined interests in which the state secures peace and order and refrains from interfering in the competitive activity of private appropriation. This clearly contrasts with the neo-Malthusian theory of constraint of activity to degrees and extent determined by ecological exigencies.

If the effect of expansionist problematic is a theory of a crisis-free ecology, this expansionist problematic, itself, is the effect of a liberal Hobbesian anthropology. As in our consideration of the neo-Malthusian problematic, the question of what cause is this anthropological premise an effect re-emerges here. But once again, before answering the question, let us await the completion of our summary of the other theories of ecology in terms of their theoretical problematic and premises.

ECO-DEVELOPMENT: THEORETICAL SPECIFICATION OF OBJECT AND THEORETICAL PREMISE

The problematic of eco-development poses ecology or rather the state of ecology as the effect of a relation between human activity and physical nature. The relation is posed as one of 'consumption of nature'. In this relation, production is collapsed into consumption as one undifferentiated moment such that the productive activity of 'working' nature into products destined for consumption is at the same time consumption of nature. The problematic poses this relation of consumption of nature as a technical relation, i.e., form of consuming nature, or technology. Indeed, ecological disorder is posed as the effect of particular form(s) of consuming nature. This problematic poses social inequality in the same way. The present form(s) of consuming nature has (have) an injurious effect on ecology and an unjust effect on social structure. Ecological disorder and social inequality are posed, thus, in terms of present technologies – the form of technology of the rich being destructive of nature one way and that of the poor another way, both forms of which reproduce a poor/rich dichotomy.

Unlike the problematic of neo-Malthusianism which defines ecological crisis in purely quantitative terms as excesses, this problematic poses ecological crisis as effects of technology and excesses derive from particular technologies. The rich are excessive in one way, the poor another in virtue of their respective technologies. Hence, the rich are rich and the poor are poor by virtue of their direct relation to nature, i.e. consumption of nature or technology, and not by virtue of a relation to one another.

If the problematic poses questions of poverty and ecological disorder as effects of technology, it invites a solution to such problems of rich/poor dichotomy and ecological crisis in similar terms. The solution is eco-technology posed as self-reliance of local community. It is a technology which is defined in terms of basic human needs of local community and ecological exigencies both of which are posited as moderators of technology and hence as regulators of consumption of nature both quantitatively (limiting production/consumption to necessary production/consumption as opposed to excessive) and qualitatively (compatible with natural processes as opposed to destructive). Cultural specificity of community and ecological uniqueness of habitat as determinants of eco-technology imply a pluralism of eco-technology and also a pluralism of eco-community. Hence, the restoration of ecological balance and establishment of social equality and Third World independence takes the form of a global pluralism of eco-technology and eco-community.

The sense of community in this theory is determined by a problematic of direct consumption of nature in that it is posited solely in terms of the relation with nature as a technical labour process – i.e. production/consumption of nature mediated by eco-technology – and not in terms of collective decision-making regarding the objective of labour and the disposition of the social product. Since the problematic leaves aside as 'unproblematical' social relations, it allows a sense of community as a technical division of labour determined by eco-technology. Such a sense of community has implications for Third World struggle for autonomy from the centre. It implies that the struggle for independence consists of a learning process for technological self-reliance. Hence the struggle for independence in Third World social formations is displaced from the level of struggle for economic and political independence from the centre to the level of technological learning for self-reliance as if this learning process suffices as means to independence.

Underlying this problematic is an anthropological premise of human essence but one which differs from the Hobbesian anthropology. This anthropology is an anthropology of self-sufficiency. Although self-sufficiency is conveyed in the theory as a property or inherent essence of community, the logic of the theoretical construction which limits self-sufficiency to technology would suggest that it is not an anthropology of collective essence or collective subject but rather an aggregate of individual subjects. It is an anthropology of solitary individual self-sufficiency. It is a premise which recurs in radical ecology in a less ambiguous form. Before we examine the basis of this premise let us look again at the theoretical construction of radical ecology.

RADICAL ECOLOGY: THEORETICAL SPECIFICATION OF OBJECT AND THEORETICAL PREMISE

Radical ecology is distinguished first and foremost from neo-Malthusianism and eco-development by a problematic which advances a notion of ecological crisis as the effect of social relations. Ecological destruction as any form of despoliation of nature including the effects of multifaceted pollution, resource depletion etc. is simply a symptom of hierarchical relations. This form of relation which is seen as being pervasive in contemporary society is posed in terms of a subject–object relation and hence one of control. One party is the subject determining the conditions of life of the other party. This latter remains a mere object denied the expression of its subjectivity and subscribing (willingly or unwillingly) to the directives of the subject. The relation of nature to society is not only analogous to the relation between the socially dominated and the socially dominant but determined by the social relation of control. The problematic which links ecological crisis to hierarchical social relations invites a solution of radical change on the level of social relations. Indeed, it is a solution of restoration of free subjectivity by supplanting relations of control by equality between individuals – a solution which at the same time restores ecological balance.

Although this basic problematic is common to all radical ecology there are variations between particular theories. The subject–object relation in the problematic of Bookchin pervades all spheres of social relations with no necessary relation of determination nor of overdetermination between these spheres and disallows any historical materialist conception of domination. That is, the economic relation of private property is neither determinant nor overdeterminant, in the last instance, of social structure. The problematic posits, rather, a relation of independence between forms of domination. It is a problematic in which the subject of change is the individual. It is an individual who discards all aspects of himself as object by self-liberation from authority. It is the individual who posits objectives and develops means to achieve them without recourse to external direction nor hindrance from external authority. It is the self-sufficient individual in a radical sense of solitary self-sufficiency.

In the problematic of Gorz, the economic relation is determinant, in the last instance, of social structure. Hence the dominant and determinant subject–object relation is the economic relation between non-producers (capitalist class) and direct producers (working class). The sense of 'object' as the direct producer(s), in this relation, is not simply control by the non-producer. In this control, more importantly, lies the denial of expression of subjectivity of the direct producer. The problematic poses free subjectivity as labour such that the direct producer is subject and subject is direct producer positing objectives of his labour, developing means to achieve them without external interference or control.

The notion of labour posited is technical labour as technical expertise to complete operations from beginning to end of a labour process. The solution to the subject–object relation which the problematic calls upon is the dissolution of the relation and a restoration of free subjectivity. The technical sense of labour suggests a form of subjectivity as solitary individual self-sufficiency in which the individual in theory and practice commands the whole labour process of which he or she is the author without external, social control or hindrance.

Leiss's problematic of human needs and their satisfaction situates the subject–object relation in the sphere of exchange of the market economy. The consumer is posed as object in relation to the commodity as 'a tangible artifact produced and exchanged for the satisfaction of needs'. The consumer is object as lacking the means of determining his or her needs and realising them in the relation consumer-commodity, and hence is controlled in this relation. The control of object/consumer derives from a lack of 'craft knowledge concerning the quality of things' to judge the suitability of the commodity for his or her needs.

The notion of object carries the same sense of control as denial of expression of 'human essence' that it does in the Bookchin and Gorz constructions. The problematic invites a similar solution to that of Bookchin and Gorz, namely, the dissolution of the relation consumer/commodity as object/subject by restoring to the consumer his subjectivity in the form of 'craft knowledge' to produce means to satisfy his needs as he defines them. This subjectivity is individual self-sufficiency.

Underlying all three problematics is a common anthropological conception of human essence. It is a premise which assumes that a human essence inheres in human beings and that this essence is individual self-sufficiency. This is not the self-sufficiency of the private individual of liberalism but rather the self-sufficiency of the solitary individual who is neither a social means nor employs others as social means to achieve his or her posited objectives. The individual achieves freely posited objectives in direct interaction with nature. This 'free subjectivity' as socially unmediated interaction with nature is solitary self-sufficiency.

THE SOCIAL BASIS OF ANTHROPOLOGICAL PREMISES

If this assessment of problematics and underlying premises is tenable it reveals the limits of the corresponding theory as a theory of explanation and a theory of political action. The neo-Malthusian quantitative problematic reduces crises to numerical disproportions between excessive wants and limited physical resources leaving social relations of production outside the problematic and hence removed from crisis. Rather, an anthropology of infinite appetites and wants becomes the source of neo-Malthusian explanation of crises. It also constitutes the core of the political theory of restraint or political 'coercion' as a solution to the crisis of disproportion.

If the anthropology of 'possessive individualism' as theoretical premise is determinant of neo-Malthusian conception of ecological crisis and solution to crisis, the validity of neo-Malthusian theory depends on the soundness of this premise. It claims universal validity as if a 'possessive individualism' were inscribed in the nature of things. It is assumed that man possesses an inherent essence 'by nature' and that this essence is expressed or externalised in the activity of acquisition. That this essence is not posed in the same favourable terms as it is in expansionism does not affect the character of its claim, i.e. as being ontologically founded. The critical tone with which this essence is posed affects the modality of its expression. Rather than implying free, unconstrained expression of this essence, the critical cast implies external regulation. Hence, the expression of this essence is determined not merely by its own internal subjectivity but by a force external to it. This is the sense of the coercive state as regulator of the 'excessive wants inherent in human nature'.

A claim to universal validity does not, in itself, imply that the claim is scientifically based. Whatever the implicit or explicit claims to universal validity of this anthropological premise it has neither a universal character nor a scientific basis. Rather, it is ideological in the sense that it is unaware of its own grounds of determination, of its own presuppositions and its implications and in the sense that its relevance and validity are partial and not universal. The acquisitive activity which the anthropological premise fixes in human nature is rather inscribed in determinate social relations. Furthermore, the conception of this acquisitive activity, as activity determined by human nature, is itself the ideological effect of determinate social relations. In other words, a particular form of social relations determines this form of activity and determines as well the conception of this activity as anthropologically determined activity.

The form of these relations is not a mystery. It is the capitalist social relations. In the light of the rich and extensive historical materialist literature, this statement is banal. For many, Marx's own work provides sufficient demonstration of the ideological effect of the capitalist relations of production as necessarily an individualist anthropology. It shows how the structural separation of the sphere of circulation from that of production has the effect of limiting the field of conceptualised activity. That is, the sphere of circulation where exchange is based on equivalence becomes alone the site of conceptions of the whole capitalist structure. Accordingly, equality, freedom and individualist mastery become the core principles of the ideology of capitalism, namely liberal ideology.[2] Lukács' work on philosophical systems of thought takes up Marx's insights into the structure of capitalist social relations in arguing the thesis that the perspective of the bourgeois class mediates liberal philosophy from Kant to Hegel.[3] Another work in this line is that of C.B. Macpherson on 'possessive individualism' which relates the principal tenets of Hobbes and Locke regarding human nature to capitalist social relations as opposed to their anthropological claims of universal human essence.[4]

In the light of these and other similar studies the point that the anthropology of possessive individualism is the ideological effect of capitalist relations of production need not detain us further here. What is important for our understanding of neo-Malthusianism and for our assessment of its limits as a theory of ecological restoration is its class determination. It is both a critique and defence of capitalist relations of production. This critique/defence is mediated by the perspective of the bourgeois class. The conception of human nature as infinitely acquisitive is a bourgeois conception which in neo-Malthusianism is assumed to be universal. This assumption determines the parameters of the critique of expansionary activity. It is a critique of quantity not of the capitalist relations of production themselves. Indeed, we see in the neo-Malthusian critique of capital accumulation, which covers as well public capital of socialist systems, a separation of capital expansion from relations of private property. This assumed independence of capital accumulation from relations of private property invites a critique of the former as excessive and a defence of the latter as expressive of human nature. The solution it invites is the regulation of appropriating activity at a 'steady-state' level. Of course, this solution is antithetical to the dynamic of appropriating activity, itself, as relations of private property propel capital accumulation.

The neo-Malthusian contradiction between capital expansion and relations of private property does not emerge in expansionist theory for there is no call for restraint on expansionary activity but, on the contrary, a plea for its free rein. The differences between the two hence are considerable but their common anthropological assumption unites them. The assessment of social activity and ecology is cast in both cases from the bourgeois class perspective, the differences being that neo-Malthusianism is a critical assessment and expansionism is not.

Radical ecology and to some extent eco-development depart dramatically from the liberal anthropological assumption of neo-Malthusianism and expansionism. The anthropological premise is one of individual self-sufficiency. This premise is the ground of a radical critique of capitalist productive relations as well as socialist productive relations in reference to concrete socialist social formations. As we saw, this critique traces the ecological crisis to such social relations and hence proposes their dissolution and replacement by individual self-sufficiency.

Just as the liberal anthropological premise is determinant of neo-Malthusian critique and revealing of its class character as bourgeois, so is the radical anthropological premise of self-sufficiency determinant of radical ecology and revealing of its class character as petit bourgeois.[5] The discussion of this question in Chapter 4 need not be repeated here. What is important to emphasise is that the self-sufficiency of radical ecology is the theoretical core of a mode of production of independent producers and the conception of the libertarian society is one in which this mode of production either prevails or is at least the dominant mode.

The struggle for equality and ecological restoration and harmony of radical ecology is the struggle of independent producers to create or establish and maintain independence. It is an economic form of struggle from capitalist exploitation and domination (as well as socialist domination) and ecological despoliation.

Determining the sense in which this form of struggle is petit bourgeois class determined does not, in itself, constitute an indictment of radical ecology. Nor does the fact that the mode of production of small or independent producers ever exist historically as a dominant mode of production suffice to challenge radical ecology. These factors, however, do help situate the limitations of radical ecology as a programme of social change. On a purely strategic level it is highly doubtful that the capitalist mode of production can be seriously challenged by a form of struggle in which individuals or groups of individuals place themselves outside market relations by developing means of livelihood including food production, lodging, health care and so on. In fact, this is a practical option only for some individuals and some groups of individuals. But were such individuals and groups of individuals more numerous and were the scale of such self-sufficiency more extensive, market relations would not simply dissipate. Escapism or withdrawal is hardly the solution to the problem.

TOWARDS A POLITICAL ECONOMY THEORY OF ECOLOGY

Outlining the limits of these forms of ecological critique – the neo-Malthusian, the critique of eco-development and of radical ecology – does not imply that these forms of critique are individually and collectively void of any merits. Indeed, their principal and far from insignificant contribution is the attention they draw to ecology. The present state of ecology is raised as a global social issue demanding an immediate and comprehensive response. They point to the alarming degrees of natural despoliation as a tragic fact of our time. Regardless of the perspective from which this ecological destruction is understood, the emphasis on this very issue is, in itself, an important contribution to raising the awareness of the contemporary ecological problems as being linked in some way to social activity.

All forms of critique draw a relation between accumulation and ecological despoliation. The theoretical limitations of each of these critical perspectives on ecology do not blur this critical relation but on the contrary set it in focus as the core of the problem. In spite of the widely disparate assessments of accumulation from the neo-Malthusians to the radical ecologists, ecological despoliation is generally regarded to be its effect. There is no dispute about this point at least. One may be tempted to say that even a common sense observation could not fail to relate ecological despoliation to accumulation.

If accumulation is the source of the present ecological problem it is imperative that the dynamics of this source be understood if the intent is the search of a solution to the ecological crisis. Whatever may be the objections to radical ecology, its contention that the source of accumulation is social relations must be retained as the starting point to an understanding of this issue. Much of its critique is based on those social relations of capitalist accumulation and the essential factor of such social relations is that capital accumulation is private. The working class produces and the non-producing class appropriates the surplus product privately as its own private property. It is this relation which is at the heart not only of the present ecological crisis but of economic crisis including the economic and political marginalisation of the Third World. The source of the crisis lies not in the production of a surplus product as such in spite of overtones in the radical ecological arguments to this effect. Indeed, it is difficult to conceive of a modern society without a socially produced surplus product disposed for hospitals, museums, schools, roads and so on. The source of the crisis rather lies in the privatisation of the surplus product as private property. In the relation of private property lies the propulsion for expansion of profitable commodity production. Leiss is correct about the ubiquity of the commodity even though he does not link it to productive relations. Whatever may be made to have an exchange value from aids to brushing mushrooms to instruments for brushing toe-nails enters into the circuit of commodity production. What Leiss ignores as we saw, is that labour-power is also a commodity having an exchange-value and use-value as does any other commodity. There is, however, an essential difference in this commodity. It is a difference of no mere academic interest but of political interest for it contains the key to adequate action for social change and ecological restoration.

The commoditisation of labour-power is an effect of the privatisation of the means of production. The separation of direct producer(s) from the means of production historically and structurally marked by capitalist productive relations engenders the exchange of the labour-power of producers for a salary which, in addition to representing the means of reproduction of labour-power, implies the possibility of access to the means of work. The elements of this exchange are too well-known to require a lengthy overview here. We may simply recall that the labour-power in use (i.e. working to produce commodities) generates a value in excess of the value (payment) for which it is exchanged. This excess takes the form of private capital. In these productive relations the owners of the means of production seek to increase this excess and the producers to reduce it and herein lies the economic struggle of the opposing classes.

This brief reference to political economy is not to divert attention away from ecology but on the contrary to insist on their interconnection. Ecological destruction, as we know it in our capitalist social formations, is an effect of this political economy. But the sense of this interconnection varies in relation to the perspective from which it is assessed. We have attempted to demonstrate in this essay that the sense which each form of ecology draws of this interconnection varies in relation to the perspective from which it is understood. Further, that this perspective is not a classless perspective totally ecaping the ideological effects of class structure. Rather, if our interpretation is correct, neo-Malthusianism owes its coherence largely to the bourgeois class perspective and eco-development and radical ecology to a petit bourgeois class perspective. If these perspectives lend an evident coherence to these forms of ecological thought they also account for limitations in the grasp of the interconnection between political economy and ecology. Lukács's insight into the relationship between understanding and class perspective constitutes a theory of knowledge which casts much light on that class perspective which allows a more global grasp of political economy and the contradictions inherent in capitalist relations.[6] This is the perspective of the working class which brings into better focus ecological destruction as the effect of exploitative productive relations. An integration of political economy into ecological theory would provide that perspective.[7]

A political economy theory of ecology would situate the working class struggle at the forefront of the political struggle for economic, political and social change and ecological restoration without providing a magic formula containing an instant success of this struggle in achieving social and ecological liberation. It would emphasise that the struggle for change in those social relations whose effects are socially exploitative and repressive and ecologically destructive must ultimately involve that class in the relation whose labour-power, once sold and beyond its control is put to socially exploitative and ecologically destructive ends. That class must be an integral part of the ecological struggle. It must acquire control of its own labour-power and determine the ends to which it is put. A fundamental change at the level of relations of production must comprise this. Only then could activity as individual expression in determining conditions of life be meaningful – activity which may bear some characteristics of self-sufficiency.

On the level of political struggle for environmental protection the labour movement is far from absent. No one is more aware of this than radical ecologists, particularly Robert Paehlke, who has provided a great deal of evidence of the considerable priority which the North American labour movement accords to such ecologically-related issues as occupational health and safety. He has demonstrated as well, the affinity between these concerns of labour and those of the environmental movement and of a growing cooperation between the environmental and labour movements.[8] Indeed, some important reform legislation for greater safety protection owes to a joint pressure on the state from these two movements.

The issue is not whether or not there is an affinity between ecological concerns and concerns for the social good of the dominated classes. To most, there is little doubt of this, least of all to radical ecologists. That ecological restoration must be inextricably tied to change of the productive relations of exploitation is clear.

What is not clear, particularly to Gorz, is that the struggle for ecological restoration in our own social formations is necessarily tied to the working class struggle for liberation from capitalist productive relations. This is the crucial point. The claim of radical ecology is that this social change can occur by action of self-sufficiency. This is posed as a condition for change as the form of transition or revolutionary action and as the form of activity constitutive of the free or libertarian society. But this claim for action, as we saw, points to a path which only the solitary individual can tread. It points to a path leading to the decomposition of society rather than to a free society. Can it be seriously entertained as a substitute for collective, class action for change from exploitative productive relations without the risk of its becoming simply separate from but coexistant with the dominant social structure of exploitation and domination?

Is there not a displacement of the ecological issue from the level of social relations to the level of technology in the forms of socio-ecological theory we have examined? This is surely the case for eco-development and for a number of ecologists who point to technology as the source of environmental destruction.[9] We saw how the general critique of radical ecology including the critique of social relations accords technology determining properties, a factor entirely consistent with the premise of asocial self-sufficiency. We saw how this premise motivates not only the critique of present forms of technology conceived as being endemic to the ecological crisis but also serves to pose questions of liberation from domination in terms of a new form of technology, namely a craft technology as one of a form and scale which the individual alone can command.

It cannot be denied that modern technology is a contributory factor in ecological destruction and that there is a dire need for changes in some forms of technology such as nuclear technology. But the link of determination between technology and social relations must be grasped in such a way that the primacy of social relations over technology and not technology over social relations may suggest directions for social and ecological liberation. This is the direction Barry Commoner adopts in his final chapter of *Poverty of Power*, clearly distinguished from his earlier *The Closing Circle*, in which he situates technology in its relation to the capitalist relations of production and lays some of the ground work for a political economy theory of ecology. His critique of technology is not simply a critique of faulty productive processes but rather reveals the inherent contradictions of capitalist relations of production as expressed in the tendency of the rate of profit to fall. As Commoner points out, the process of incorporation of new technology in the productive process to increase productivity and to counteract this tendency is a process which is not only ultimately incapable of resisting the fall of the rate of profit but also is the root of environmental destruction. In Commoner's analysis there is an implicit and explicit reference to the law of value in accounting for the source of value in labour and hence in pointing to the opposing processes of expansion of constant capital (incorporation of new technological innovation to increase productivity and reduce costs) on the one hand, and reduction of variable capital (labour) on the other, as contradictory. Rather than stabilising or increasing the rate of profit, they are the source of its falling tendency and of economic crises.

Commoner not only paves the road to a political economy theory of ecology but draws political conclusions from such a perspective which lead the way to socialism as an ecologically sound and socially just form of social change. For those who identify socialism with centralisation, he adds a concluding note of caution against rejecting it.

> It seems unrealistic . . . at this moment in history, to categorically reject a socialist economy on the grounds that its political form is necessarily repressive and therefore abhorrent to . . . democratic freedoms . . . That no existing example of a socialist society – whether the USSR, China, or Cuba – is consistent with the political democracy inherent in US tradition means that wholly new political forms would need to be created.[10]

He does not work out a political programme for this, however, nor does he try to do so, nor can any one theorist hope to do so. That is worked out on the level of political praxis in which the working class must be a critical force.

Notes and References

Introduction

1. The most notable exponent of this view is Amory Lovins. See, in particular, his *World Energy Strategies: Facts, Issues, and Options* (Ballinger Publishers, Cambridge, 1975); and in collaboration with John H. Price, *Non-Nuclear Futures: The Case for an Ethical Energy Strategy* (Ballinger Publishers, Cambridge, 1975). Countless others add to this critique of nuclear energy. B.L. Welch, 'Nuclear Power Risks: Challenge to the Credibility of Science', *International Journal of Health Services* vol. 10, no. 1, 1980, pp. 5–36; J.M. Brunet, *Les dangers de l'énergie nucléaire* (Éditions du Jour, Montréal, 1977); Walter Patterson, *Nuclear Power* (Penguin Books, Markham, Ont., 1976); John Rensenbrink, 'The Anti-Nuclear Phenomenon: A New Look at Fundamental Human Interest', *New Political Science* no. 7, Fall 1981, pp. 75–89.
2. Such as members, supporters and sympathisers of Friends of the Earth.
3. This is the effect that Timothy Luke's theory of a 'deconstructionist' ecology would wish to have in practice. See his 'Notes for a Deconstructionist Ecology' in *New Political Science*, no. 11, Spring 1983, pp. 21–32.
4. Murray Bookchin, *Toward an Ecological Society* (Black Rose Books, Montréal, 1980).
5. One of the most notable is Murray Bookchin and in chapter 4 we look at his work in some detail.
6. André Gorz is one of the principal theorists in this tradition and in chapter 5 we examine his work and note the theoretical affinity with that of Bookchin.

1 Neo-Malthusian Theory

1. Thomas Malthus, *An Essay on the Principle of Population as it Affects the Future Improvement of Societies, with Remarks on the Speculations of Mr. Godwin, M. Condorat, and other writers*, 1798 edition (reprints of Economic Classics, N.Y., 1965).
2. Thomas Malthus, *An Essay on the Principle of Population or a View of its Past and Present Effects on Human Happiness with an Inquiry into our Prospects Respecting the Future Removal or Mitigation of the Evil which it Occasions*, 1801 edition (Richard D. Irwing, Homewood, Illinois, 1963).
3. Paul Ehrlich, *The Population Bomb* (Ballantine, N.Y., 1968); in collaboration with Anne H. Ehrlich, *Population, Resources, Environment: Issues in Human Ecology*, 2nd Edition (W. H. Freeman and Co., San Francisco, 1972); and in collaboration with Dennis Pirages, *ARK II: Social Response to Environmental Imperatives* (W. H. Freeman and Co., San Francisco, 1974).

4. Garrett Hardin, 'The Tragedy of the Commons', in *Science*, Vol. 162, Dec. 1968, pp. 1243–8; *Population, Evolution and Birth Control; A Collage of Controversial Ideas* (W.H. Freeman and Co., San Francisco, 1969).
5. Georg Borgstrom, *The Hungry Planet* (Collier-Macmillan Ltd., London, 1965); *Too Many* (Collier-Macmillan Ltd., London, 1969).
6. The Club of Rome is an association of scientists of various disciplines formed in 1968 on the initiative of Aurelio Peccei. Since its inception it has consistently cultivated a non-partisan, universalist image. It refrains from too close an association with any given state government or group of allied states, as opposed to others and it promotes its neo-Malthusian theory as one founded on objective laws of nature and society unmediated by private, partisan interests. As an independent international organisation chartered under the laws of Switzerland with headquarters in Geneva, the world capital of international organisations, it claims independence from formal political ties with any particular state or group of states placing itself on the level of universalistic interests and above particular interests of given state governments. As an association of a select membership of eminent scientists from all parts of the world, none of whom has formal affiliations with state governments, it claims a neutral approach to economic and social development and one which yields the correct scientific solution to the crisis of overconsumption of nature. The Club defends its claim to scientific objectivity as non-partisan neutrality in its endorsement of the scientific reports submitted to it on the world economic and social condition and prepared by club members and sympathisers. The first report, *The Limits to Growth*, was prepared by a group of seventeen scientists including American, German, Indian, Turkish, and Norwegian collaborators under the leadership of Donella and Dennis Meadows of the Massachusetts Institute of Technology and known as the MIT Team. Other reports follow, each authored by a different team of scientists. Amongst these are: Mihajlo Mesarovic and Eduard Pestel, *Mankind at the Turning Point* (The Second Report to the Club of Rome) (The New American Library, Inc., N.Y., 1974); Antony J. Dolman (ed.), *RIO, Reshaping the International Order* (The New American Library, Inc., N.Y., 1977); Ervin Laszlo *et al.*, *Goals for Mankind* (E.P. Dutton, N.Y., 1977); James W. Botkin *et al.*, *No Limits to Learning* (Pergamon Press, N.Y., 1979); Thierry de Montbrial, *Energy: The Countdown* (Pergamon Press, N.Y., 1979); Orio Giarini, *Dialogue on Wealth and Welfare* (Pergamon Press, N.Y., 1980). Each shares basic neo-Malthusian premises and each proposes a course of economic, political and social action on the basis of the universal reason of science. The value neutrality tenet, of course, is not without considerable contestation. See, for example, Jürgen Habermas, 'Between Philosophy and Science: Marxism as Critique', in *Theory and Practice* (Heinemann, London, 1974), pp. 195–252; 'The Analytical Theory of Science and Dialectics', in *The Positivist Dispute in German Sociology*, Theodor Adorno *et al.*, (Heinemann, London, 1976) pp. 131–162; Albrecht Wellmer, *Critical Theory of Society* (The Seabury Press, N.Y., 1974); Roberto Miguelez,

Science, Valeurs et Rationalité (Editions de l'Université d'Ottawa, Ottawa, 1984).

7. Donella Meadows *et al.*, *The Limits to Growth* (The New American Library Inc., N.Y., 1972).
8. The model is that of J.W. Forrester posed in *Industrial Dynamics* (MIT Press, Cambridge, Mass., 1961). This model and the entire MIT study based on this model has received much critique. See in particular *Models of Doom: A Critique of the Limits of Growth*, H.S.D. Cole *et al.* (eds) (Universe Books, N.Y., 1973).
9. This study is based on World demographic estimates for the last 3 centuries, for example, drawn from A.M. Curr-Saunders, *World Population: Past Growth and Present Trends* (Clarendon Press, Oxford, 1936); Donald Bogue, *Principles of Demography* (John Wiley and Sons, N.Y., 1969); *World Population Data Sheet 1968*, Population Reference Bureau, Washington, 1968.
10. William Ophuls, *Ecology and the Politics of Scarcity: Prologue to a Political Theory of the Steady State* (W.H. Freeman and Co., San Francisco, 1977).
11. Donella Meadows *et al.*, *The Limits to Growth*, p. 46.
12. Ibid.
13. Ibid., p. 48.
14. Ibid., p. 49.
15. Williams Ophuls, *Ecology and the Politics of Scarcity*, p. 60.
16. Ibid., p. 113.
17. Ibid., p. 127.
18. Ibid., p. 130.
19. See, in particular, C.B. Macpherson, *The Theory of Possessive Individualism: Hobbes to Locke* (Oxford University Press, Oxford, 1962).
20. Garrett Hardin, 'The Tragedy of the Commons', in *Science*, Vol. 162, Dec. 1968, pp. 1243–8.
21. J. Fletcher, *Situation Ethics* (Westminster, Philadelphia, 1966).
22. Garrett Hardin, 'The Tragedy of the Commons', p. 1245.
23. Ibid., p. 1245.
24. Ibid., p. 1246.
25. William Ophuls, *Ecology and the Politics of Scarcity*, p.165.
26. See, for example, Paul Q. Hirst, 'Economic Classes and Politics' in A. Hunt (ed.), *Class and Class Structure* (Lawrence and Wishart, London, 1977); Nicos Poulantzas, *Political Power and Social Classes* (New Left Books, London, 1973).
27. William Ophuls, *Ecology and the Politics of Scarcity*, p. 226.
28. Ibid., p. 156.
29. Kenneth Boulding, 'The Economics of the Coming Spaceship Earth' in *Toward a Steady-State Economy*, Herman Daly (ed.), (W.H. Freeman & Co., San Francisco, 1973), pp. 121–32.
30. Herman Daly, 'The Steady-State Economy: Toward a Political Economy of Biophysical Equilibrium and Moral Growth' in *Toward a Steady-State Economy*, p. 152.
31. Ibid., p. 157.
32. Ibid., pp. 157–8.

33. Ibid., p. 169.
34. Ratko Milisavljevic, *Environment, idéologie et science* (Éditions anthropos, Paris, 1978).
35. Ibid., p. 315. (This and all subsequent translations of French texts are my own).
36. Ibid., p. 396.
37. Ibid., pp. 388–389. (author's emphasis).
38. Ibid., p. 403.
39. Lester Brown, *World without Borders* (Random House, 1972), p. 351.
40. Harold and Margaret Sprout, *Towards a Politics of the Planet Earth* (Nostrand Rinehold Co., N.Y., 1971), p. 476.
41. Aurelio Peccei, *One Hundred Pages for the Future, Reflections of the President of the Club of Rome* (Pergamon Press, N.Y., 1981), pp. 110–11.
42. Ibid., p. 113.
43. Lester Brown, *World without Borders*, pp. 225–6.
44. C.B. Macpherson, *The Political Theory of Possessive Individualism*.
45. Karl Marx, *Capital* (International Publishers, N.Y., 1967), particularly Vol. 1, ch.6.

2 Expansionism

1. Herman Kahn, *World Economic Development: 1979 and Beyond* (Westview Press, Boulder, Colorado, 1979), p. 23.
2. Miguel A. Ozorio de Almeida, 'The Confrontation Between Problems of Development and Environment', in *International Conciliation*, January 1972, no. 586, p. 43.
3. Herman Kahn and Anthony J. Wiener, *The Year 2,000: A Frame for Speculation on the Next Thirty-Three Years* (Collier-Macmillan, London, 1967), p. 116.
4. Herman Kahn, *World Economic Development: 1979 and Beyond*.
5. John Maddox, *The Doomsday Syndrome* (Macmillan, London, 1972), pp. 21–2.
6. Ibid., pp. 65–6.
7. Ibid., p. 78.
8. Wilfred Beckerman, *In Defence of Economic Growth* (Jonathan Cape, London, 1974), p. 218.
9. John Maddox, *The Doomsday Syndrome*, p. 88.
10. Ibid., p. 91.
11. Wilfred Beckerman, *In Defence of Economic Growth*, pp. 227–8.
12. Cited in *ibid.*, pp. 229–30
13. Ibid., p. 229.
14. John Maddox, 'The Numbers Game', in *The Doomsday Syndrome*, pp. 29–61.
15. Ibid., p. 63.
16. Ibid., pp. 195–214.
17. Herman Kahn, *World Economic Development*, pp. 74–5.
18. Ibid., pp. 60–1.
19. Ibid., pp. 163–4.

20. Ibid., p. 73.
21. John Maddox, *The Doomsday Syndrome*, p. 72.
22. Wilfred Beckerman, *In Defence of Economic Growth*, p. 141.

3 Theory of Eco-development

1. Howard E. Daugherty, Charles A. Jeanneret-Grosjean, H.F. Fletcher, *Ecodevelopment and International Cooperation, Joint Project on Environment and Development 6*, Environment Canada, CIDA, Ottawa, 1979, p. 1.
2. Ibid.
3. Ignacy Sachs, *Stratégies de l'écodéveloppement* (Editions économie humanisme, les éditions ouvrières, Paris, 1980), p. 22. (This and all subsequent translations are my own).
4. Ibid., p. 25.
5. *The Cocoyoc Declaration*, Cocoyoc, Mexico, October, 1974, p. 4.
6. Johann Galtung, 'The Basic Needs Approach', in *Human Needs*, Katrin Lederer (ed.) (Oelgeschlager, Cambridge, Mass., 1980), pp. 55–126.
7. Ignacy Sachs, *Stratégies de l'écodéveloppement*, p. 49.
8. Ibid., p. 58.
9. Ignacy Sachs *et al.*, *Initiation à l'écodéveloppement*, (Toulouse, Privat, 1981), p. 26.
10. Ignacy Sachs, *Stratégies de l'écodéveloppement*, pp. 127–8 (author's emphasis).
11. E.F. Schumacher, *Small is Beautiful, A Study of Economics as if People Mattered* (Blond & Briggs, London, 1973).

4 Theory of Eco-anarchism: Bookchin's Critique of Authority

1. Murray Bookchin, *The Ecology of Freedom: The Emergence and Dissociation of Hierarchy* (Cheshire Books Palo Alto, 1982), p. 73.
2. Murray Bookchin, *Post-Scarcity Anarchism* (Black Rose Press, Montréal, 1971), p. 285.
3. Murray Bookchin, *Toward an Ecological Society* (Black Rose Press, Montréal, 1980), p. 189.
4. Murray Bookchin, *The Limits of the City* (Harper & Row, N.Y., 1974), p. 126.
5. Murray Bookchin, *The Ecology of Freedom*, p. 70.
6. Ibid., p. 69.
7. The elimination of error by means of the process of deductively derived experimental falsification of hypotheses is the path to scientific discovery proposed by Karl Popper. See, in particular, his *Logic of Scientific Discovery* (Hutchinson, London, 1958).
8. Jürgen Habermas, 'Technology and Science as "Ideology"', in *Toward a Rational Society* (Beacon Press, Boston, 1970), pp. 81–122.
9. Murray Bookchin, *Post-Scarcity Anarchism*, pp. 45–6.
10. Ibid.
11. Ibid., p. 221.
12. Murray Bookchin, *The Limits of the City*, p. 127.

13. Murray Bookchin, *Post-Scarcity Anarchism*, pp. 59–60.
14. Ibid., p. 71.
15. Murray Bookchin, *Toward an Ecological Society*, p. 93.
16. Murray Bookchin, *Post-Scarcity Anarchism*, p. 285.
17. Ibid., pp. 138–9.
18. See C.B. Macpherson on this question. *The Political Theory of Possessive Individualism: Hobbes to Locke*.
19. Nicos Poulantzas, *Les classes sociales dans le capitalisme aujourd'hui* (Editions du Seuil, Paris, 1974). See also Christian Baudelot and Roger Establet, *L'école capitaliste en France* (Maspero, Paris, 1980).
20. In Jürgen Habermas's analysis of this question we find a defence of the thesis according to which intellectual labour comprises a part of the working class as indirectly productive of surplus-value. His analysis demonstrates, however, that the capitalist relations of production determine the productive forces and that science and technology become ideology in the capitalist relations. See, 'Science and Technology as "Ideology"', in *Toward a Rational Society*. See also his *Legitimation Crisis* (Beacon Press, Boston, 1975).
21. There are historical examples of anarchist movements of independent direct producers. They are essentially movements of resistance to usurpation by encroaching capital and to loss of economic independence.
22. Of course, as mentioned before, these workers may and do join forces with the working class through trade unions and adopt working class positions in radical action, but the form of radicalism we are identifying here is specifically petit bourgeois radicalism.

5 Ecology According to Gorz

1. One of the best exponents and proponents of humanist Marxism is Georg Lukacs. See, for example, *The Ontology of Social Being*, Vol. 3, *Labour*, trans. David Fernbach (Merlin Press, London, 1980).
2. André Gorz, *Le Socialisme difficile* (Éditions du Seuil, Paris, 1967), p. 62.
3. Gorz claims that while automation has not entirely replaced manufacture, it is characteristic of 'hi-tech' industries such as the nuclear, the petrochemical, heavy machinery, etc., which are dominant in the economy and lead the trend toward generalised automation.
4. André Gorz, *Stratégie ouvrière et néocapitalisme* (Éditions du Seuil, Paris, 1964), p. 105.
5. Ibid.
6. Ibid., p. 106.
7. Gorz cites the case of the Neyrpic and Bull-Gamgetta conflicts of the early '60s in *Le Socialisme difficile*, p. 65.
8. André Gorz, *Critique de la division du travail* (Éditions du Seuil, Paris, 1973), p. 254.
9. Ibid., p. 263.
10. Ibid., p. 254.
11. Ibid.

12. Ibid., p. 266.
13. Ibid.
14. André Gorz, *Farewell to the Working Class: An Essay on Post-Industrial Socialism* (Pluto Press, London, 1982), p. 29.
15. Ibid., p. 44.
16. Ibid., p. 67.
17. André Gorz, *Ecology as Politics* (Black Rose Press, Montreal, 1980), p. 25.
18. Ibid., p. 26.
19. Ibid., p. 31.
20. André Gorz, *Farewell to the Working Class*, p. 97.
21. Ibid., p. 99.
22. Ibid., pp. 100–1.
23. Ibid., p. 117.

6 Leiss's Critical Theory of Human Needs

1. William Leiss, *The Limits to Satisfaction* (University of Toronto Press, Toronto, 1976), p. 51.
2. Ibid., p. 65.
3. Ibid., p. 51.
4. Ibid.
5. Ibid., pp. 51–2.
6. Ibid., p. 64.
7. Ibid.
8. Ibid., p. 65.
9. Ibid., pp. 65–6.
10. Ibid., p. 65.
11. Ibid., pp. 51–2.
12. Ibid., p. 72.
13. Ibid.
14. Ibid.
15. Ibid., p. 73.
16. Ibid.
17. Ibid.
18. Ibid., p. 74.
19. Ibid.
20. Ibid., p. 75.
21. Karl Marx, *Grundrisse* (Penguin Books, Baltimore, 1973), p. 156.
22. William Leiss, *The Limits to Satisfaction*, p. 77.
23. Ibid., p. 137 Note 24.
24. Karl Marx, *Grundrisse*, p. 157.
25. William Leiss, *The Limits to Satisfaction*, p. 82.
26. Ibid., p. 25.
27. Ibid.
28. Ibid., p. 82.
29. Ibid.
30. Ibid., p. 85.

31. Ibid., p. 111.
32. Ibid., p. 102.
33. Ibid., p. 25.

Conclusion

1. Richard Neuhaus, for example, in perceiving environmentalism as a diversion or substitute for more pressing social questions fails to see the interconnectedness of environmental and social issues. *In Defense of People: Ecology and the Seduction of Radicalism* (Collier-Macmillan, Toronto, 1971).
2. Karl Marx, *Capital* (International Publishers, N.Y., 1967), particularly Vol. 1, ch. 6; *Grundrisse* (Penguin Books, Baltimore 1973).
3. Georg Lukács, *History and Class Consciousness* (Merlin Press, London, 1968).
4. C.B. Macpherson, *The Political Theory of Possessive Individualism* (Oxford University Press, Oxford, 1962).
5. Is it necessary to point out that our theoretical derivation of the class character of solitary self-sufficiency is far from equivalent to a common critique of environmentalism based on an empirically drawn association between middle-class status according to socio-economic indices such as level of income, place of residence, life style, etc., and adherence to environmentalism? Indeed, it is difficult to see in this critique how middle class, in itself, can be an indictment of environmentalism and vice versa on the basis of this superficially drawn association. An example of this reasoning is Irving Horowitz's 'The Environmental Cleavage: Social Ecology versus Political Economy', in *Social Theory and Practice*, vol. 2, no. 1, 1972, pp. 125–34.
6. Georg Lukács, see especially 'Reification and the Consciousness of the Proletariat' (II Antimonies of Bourgeois Thought), in *History and Class Consciousness*, pp. 110–49.
7. An early article by Hans-Magnus Enzensberger, 'A Critique of Political Ecology' *New Left Review*, Vol. 84, March/April 1974, pp. 3–31, makes this argument quite strongly. Some of its effectiveness is lost, however, in its failure to distinguish between various forms of ecological thought. It has unfortunately given rise to the implication that the political left identifies ecological theory with the neo-Malthusianism of Paul Ehrlich, for example, or that neo-Malthusianism is indistinguishable from radical ecology.
8. See, for example, his 'Occupational Health Policy in Canada' in *Ecology versus Politics in Canada*, William Leiss (ed.) (University of Toronto Press, 1979), pp. 97–129, 'Environmentalisme et Syndicalisme au Canada anglais et aux Etats-Unis', *Sociologie et Société*, vol. 13, April 1981, pp. 161–79; 'Environmentalism and the Left in North America, a Comment' in *Studies in Political Economy*, vol. 16, 1985, pp. 143–51. He is not alone in pointing to the association between the labour and the ecological movements. See also Michael McClosky, *Labour and*

Environmentalism: Movements that Should Work Together (W.H. Freeman and Co., San Francisco, 1973).

9. E.F. Schumacher, *Small is Beautiful* (Blond & Briggs, London, 1973); and Barry Commoner's earlier work such as *The Closing Circle* (Alfred A. Knopf, New York, 1971), which had much influence on ecological thought, are examples.
10. Barry Commoner, *The Poverty of Power, Energy and the Economic Crisis* (Alfred A. Knopf, New York, 1976), p. 262.

Bibliography

BEAUDELOT, CHRISTIAN and ESTABLET, ROGER, *L'école capitaliste en France* (Maspéro, Paris, 1980).
BECKERMAN, WILFRED, *In Defence of Economic Growth* (Jonathan Cape, London, 1974).
BOGUE, DONALD J., *Principles of Demography* (John Wiley & Sons, New York, 1969).
BOOKCHIN, MURRAY, *Post-Scarcity Anarchism* (Black Rose Press, Montréal, 1971).
BOOKCHIN, MURRAY, *The Limits of the City* (Harper & Row, New York, 1974).
BOOKCHIN, MURRAY, *Toward an Ecological Society* (Black Rose Press, Montréal, 1980).
BOOKCHIN, MURRAY, *The Ecology of Freedom: The Emergence and Dissolution of Hierarchy* (Cheshire Books, Palo Alto, 1982).
BORGSTROM, GEORG, *Too Many* (Collier-Macmillan, London, 1969).
BORGSTROM, GEORG, *The Hungry Planet* (Collier-Macmillan, London, 1965).
BOTKIN, JAMES W., *et al.*, *No Limits to Learning* (Pergamon Press, New York, 1979).
BOULDING, KENNETH, 'The Economics of the Coming Spaceship Earth', in Daly, Herman (ed.), *Toward a Steady-State Economy* (W.H. Freeman & Co., San Francisco, 1973).
BROWN, LESTER R., *World Without Borders* (Random House, New York, 1972).
BRUNET, JEAN-MARC, *Les dangers de l'énergie nucléaire* (Editions du Jour, Montréal, 1977).
CLARK, JOHN, *The Anarchist Moment, Reflections on Culture, Nature and Power* (Black Rose Press, Montréal, 1984).
COLE, H.S.D., *et al.*, *Models of Doom: A Critique of the Limits of Growth* (Universe Books, New York, 1973).
COMMONER, BARRY, *The Closing Circle* (Alfred A. Knopf, New York, 1971).
COMMONER, BARRY, *The Poverty of Power, Energy and the Economic Crisis* (Alfred A. Knopf, New York, 1976).
CURR-SAUNDERS, A.M., *World Population: Past Growth and Present Trends* (Clarendon Press, Oxford, 1936).
DALY, HERMAN E., 'The Steady-State Economy: Toward a Political Economy of Biophysical Equilibrium and Moral Growth', in DALY, HERMAN (ed.), *Toward a Steady-State Economy* (W.H. Freeman & Co., San Francisco, 1973), pp. 149–174.
DAUGHERTY, HOWARD E., JEANNERET-GROSJEAN, CHARLES, A., and FLETCHER, H.F., *Ecodevelopment and International Cooperation*, Joint Project on Environment and Development 6 (Environment Canada, CIDA, Ottawa, 1979).

De ALMEIDA, MIGUEL A. OZORIO, 'The Confrontation Between Problems of Development and Environment', in *International Conciliation* no. 586, Jan 1972.

De MONTBRIAL, THIERRY, *et al.*, *Energy: The Countdown* (Pergamon Press, New York, 1979).

DOLMAN, ANTONY J. (ed.), *RIO, Reshaping the International Order*, (The New American Library Inc., New York, 1977).

EHRLICH, PAUL, *The Population Bomb* (Ballantine, New York, 1968).

EHRLICH, PAUL and EHRLICH, ANNE H., *Population, Resources, Environment: Issues in Human Ecology* (W.H. Freeman & Co., San Francisco, 1972), 2nd edition.

EHRLICH, PAUL R. and PIRAGES, DENNIS, *ARK II: Social Response to Environmental Imperatives* (W.H. Freeman & Co., San Francisco, 1974).

ENZENSBERGER, HANS MAGNUS, 'A Critique of Political Ecology', in *New Left Review*, no. 84, March/April 1974, pp. 3–31.

FLETCHER, JOSEPH, *Situation Ethics* (The Westminster Press, Philadelphia, 1966).

FORRESTER, JAY W., *Industrial Dynamics* (MIT Press, Cambridge, Mass., 1961).

GALTUNG, JOHANN, 'The Basic Needs Approach', in LEDERER, KATRIN (ed.), *Human Needs* (Gunn & Hain Publishers Inc., Oelgeschlager, 1980).

GIARINI, ORIO, *Dialogue on Wealth and Warfare* (Pergamon Press, New York, 1980).

GORZ, ANDRE, *Stratégie ouvrière et néocapitalisme* (Editions du Seuil, Paris, 1964).

GORZ, ANDRE, *Le socialisme difficile* (Editions du Seuil, Paris, 1967).

GORZ, ANDRE, *Critique de la division du travail* (Editions du Seuil, Paris, 1973).

GORZ, ANDRE, *Ecology as Politics* (Black Rose Press, Montréal, 1980).

GORZ, ANDRE, *Farewell to the Working Class: An Essay on Post-Industrial Socialism* (Pluto Press, London, 1982).

HABERMAS, JURGEN, 'Technology and Science as "Ideology"', in *Toward a Rational Society* (Beacon Press, Boston, 1970).

HABERMAS, JURGEN, 'Between Philosophy and Science: Marxism as Critique', in *Theory and Practice* (Heinemann, London, 1974).

HABERMAS, JURGEN, *Legitimation Crisis* (Beacon Press, Boston, 1975).

HABERMAS, JURGEN, 'The Analytical Theory of Science and Dialectics', in ADORNO, THEODOR W., *et al.*, *The Positivist Dispute in German Sociology* (Heinemann, London, 1976).

HARDIN, GARRETT, 'The Tragedy of the Commons', in *Science*, vol. 162, no. 3859, Dec 1968.

HARDIN, GARRETT, *Population, Evolution and Birth Control: A Collage of Controversial Ideas*, (W.H. Freeman & Co., San Francisco, 1969).

HARTE, JOHN and SOCOLOW, ROBERT H., *Patient Earth* (Holt, Rinehart & Winston Inc., New York, 1971).

HIRST, PAUL, 'Economic Classes and Politics', in HUNT, ALAN (ed.), *Class and Class Structure* (Lawrence & Wishart, London, 1977).

HOBBES, THOMAS, *Leviathan* (ed. C.B. MACPHERSON) (Penguin Books, Markham, 1968).

HOPKINS, M.J.D., 'A Global Forecast of Absolute Poverty and Employment', in *International Labour Review*, vol. 119, no.5, Sept–Oct 1980, pp. 565–577.

HOROWITZ, IRVING LOUIS, 'The Environmental Cleavage: Social Ecology Versus Political Economy', in *Social Theory and Practice*, vol. 2, no. 1, Spring 1972, pp. 125–134.

INTERNATIONAL BANK FOR RECONSTRUCTION AND DEVELOPMENT, *Report on the Limits to Growth*, Washington, DC, Sept. 1972.

KAHN, HERMAN and WIENER, ANTHONY, J., *The Year 2000: A Framework for Speculation on the Next Thirty-Three Years* (Collier-Macmillan, *London, Ontario*, 1967).

KAHN, HERMAN, *World Economic Development: 1979 and Beyond* (Westview Press, Boulder, 1979).

LASZLO, ERVIN, *et al.*, *Goals for Mankind* (E.P. Dutton, New York, 1977).

LEISS, WILLIAM, *The Domination of Nature* (George Braziller, New York, 1972).

LEISS, WILLIAM, *The Limits to Satisfaction* (University of Toronto Press, Toronto, 1976).

LEISS, WILLIAM (ed.), *Ecology Versus Politics in Canada* (University of Toronto Press, Toronto, 1979).

LOVINS, AMORY B., *World Energy Strategies: Facts, Issues, and Options* (Ballinger Publishers, Cambridge, 1975).

LOVINS, AMORY B. and PRICE, JOHN H., *Non-Nuclear Futures: The Case for an Ethical Energy Strategy* (Ballinger Publishers, Cambridge, 1975).

LUKACS, GEORG, *History and Class Consciousness* (Merlin Press, London, 1968).

LUKACS, GEORG, *The Ontology of Social Being: Labour*, vol. 3 (The Merlin Press, London, 1980).

LUKE, TIMOTHY, 'Notes for a Deconstructionist Ecology', in *New Political Science*, no. 11, Spring 1983.

MACPHERSON, C.B., *The Political Theory of Possessive Individualism: Hobbes to Lock* (Oxford University Press, Oxford, 1962).

MADDOX, JOHN, *The Doomsday Syndrome* (Macmillan, London, 1972).

MALTHUS, THOMAS R., *An Essay on the Principle of Population or a View of its Past and Present Effects on Human Happiness with an Inquiry into our Prospects Respecting the Future Removal or Mitigation of the Evil which it Occasions*, 1801 Edition (Richard D. Irwin, Homewood, Illinois, 1963).

MALTHUS, THOMAS R., *An Essay on the Principle of Population, as it Affects the Future Improvement of Society, with Remarks on the Speculations of Mr. Godwin, M. Condorcet, and other Writers*, 1798 edition (Reprints of Economic Classics, New York, 1965).

MARX, KARL, *The Economic and Philosophic Manuscripts of 1844* (International Publishers, New York, 1964).

MARX, KARL, *Capital*, vol. 1 (International Publishers, New York, 1967).
MARX, KARL, *Grundrisse* (Penguin Books, Baltimore, 1973).
MASS, BONNIE, *Population Target: The Political Economy of Population Control in Latin America* (Charters Publishing Co., Brandon, 1976).
McCLOSKY, MICHAEL, *Labour and Environmentalism: Movements that Should Work Together* (W.H. Freeman & Co., San Francisco, 1973).
MEADOWS, DONELLA, *et al.*, *The Limits to Growth* (New American Library, New York, 1972).
MESAROVIC, MIHAJLO and PESTEL, EDUARD, *Mankind at the Turning Point* (New American Library, New York, 1974).
MIGUELEZ, ROBERTO, *Science, Valeurs et Rationalité* (*Editions de l'Université d'Ottawa*, Ottawa, 1984).
MILISAVLJEVIC, RATKO, *Environnement, idéologie et science* (Editions Anthropos, Paris, 1978).
MILL, JOHN STUART, *Principles of Political Economy* (University of Toronto Press, Toronto, 1965).
NEUHAUSS, RICHARD J., *In Defense of People: Ecology and the Seduction of Radicalism* (Macmillan, New York, 1971).
OPHULS, WILLIAM, *Ecology and the Politics of Scarcity: Prologue to a Political Theory of the Steady State* (W.H. Freeman & Co., San Francisco, 1977).
PAEHLKE, ROBERT, 'Environnementalisme et syndicalisme au Canada anglais et aux Etats-Unis', in *Sociologie et Société*, vol. 13, no. 1, April 1981, pp. 161–79.
PAEHLKE, ROBERT, 'Environmentalism and the Left in North America: A Comment', in *Studies in Political Economy* no. 16, Spring 1985, pp. 143–51.
PATTERSON, WALTER C., *Nuclear Power* (Penguin Books, Markham, 1976).
PECCEI, AURELIO, *One Hundred Pages for the Future, Reflections of the President of the Club of Rome* (Pergamon Press, New York, 1981).
PECCEI, AURELIO and SIEBKER, MANFRED, 'Problématique des limites matérielles', in *Quelles limites?* (Editions du Seuil, Paris, 1974).
PETITJEAN, ARMAND, 'La pensée des limites', in *Quelles limites?*, (Editions du Seuil, Paris, 1974).
POPPER, KARL, *Logic of Scientific Discovery* (Hutchinson, London, 1958).
POULANTZAS, NICOS, *Political Power and Social Classes* (New Left Books, London, 1973), trans. of *Pouvoir politique et classes sociales* (Maspéro, Paris, 1968).
POULANTZAS, NICOS, *Les classes sociales dans le capitalisme aujourd'hui* (Editions du Seuil, Paris, 1974).
RENSENBRINK, JOHN, 'The Anti-Nuclear Phenomenon: A New Look at Fundamental Human Interests', in *New Political Science*, vol. 2, no. 7, Fall 1981.
SACHS, IGNACY, *Environnement et développement – Nouveaux concepts pour la formulation de politiques nationales et de stratégies de coopération internationale*, Joint project on Environment and Development 2, Ottawa (Environnement Canada and ACDI, 1977).

SACHS, IGNACY, *Stratégies de l'écodéveloppement* (Editions ouvrières, Paris, 1980).

SACHS, IGNACY, *et al.*, *Initiation à l'écodéveloppement* (Privat, Toulouse, 1981).

SCHUMACHER, E.F., *Small is Beautiful, A Study of Economics as if People Mattered* (Blond & Briggs, London, 1973).

SPROUT, MARGARET and SPROUT, HAROLD, *Toward a Politics of the Planet Earth* (Van Nostrand Reinhold, New York, 1971).

THE COCOYOC DECLARATION, Cocoyoc, Mexico, Oct 1974.

WELCH, BRUCE L., 'Nuclear Power Risks: Challenge to the Credibility of Science', in *International Journal of Health Services*, vol. 10, no. 1, p. 190.

WELLMER, ALBRECHT, *Critical Theory of Society* (The Seabury Press, New York, 1974).

World Population Data Sheet 1968, Population Reference Bureau, Washington, 1968.

Index